Elementary Number Theory

Elementary Number Theory

An Algebraic Approach

Ethan D. Bolker

Bryn Mawr College

W. A. Benjamin, Inc.

1970

New York

Elementary Number Theory   *An Algebraic Approach*

Standard Book Number 8053-1018-5
Library of Congress Catalog Card Number 76-92217
AMS 1968 Classification 4065
Manufactured in the United States of America
12345M3210

The manuscript was put into production on May 28, 1969;
this volume was published on January 1, 1970

W. A. Benjamin, Inc.
New York, New York 10016

Preface

Elementary number theory is frequently taught only to those who have studied little mathematics and plan to study no more. Ambitious students learn "abstract algebra" instead, but all too often they find the axiomatic study of groups and rings sterile and irrelevant. To remedy both ills I have tried to capture in this book the excitement of my discovery that the algebra I had known for years was the perfect setting in which to recreate the traditional first theorems in number theory we owe to Fermat, Euler, and Gauss.

The exposition is tied to the study of three classical problems: the structure of the group of units of Z_n, integers representable in the form $x^2 - my^2$, and the Fermat equation $x^n + y^n = z^n$ for $n = 2, 3,$ and 4. I have concentrated on the parts of these problems in which the number theory and the algebra each serve to deepen the reader's understanding of the other. I therefore omitted topics such as continued fractions, elementary analytic number theory, and the beginnings of a general theory of quadratic forms which, though accessible to beginning students, did not lend themselves to an elementary algebraic treatment. Moreover, I have stressed the algebraic aspects of some of the traditional theorems. Wilson's theorem is derived from the unique factorization of polynomials with coefficients in a field, the structure of the group of units of Z_n from a theorem on products of cyclic groups proved for that purpose earlier in the book.

I have assumed only the algebra required to carry out these aims, less than is found in any standard course in "modern algebra." An interested

v

instructor could probably supply the necessary background and teach the course to good students with no experience in algebra. The prospective reader must know or learn the definitions of group, ring, homomorphism, equivalence relation, and quotient structure, and a few simple theorems, such as Lagrange's, which asserts that the order of a finite group is a multiple of the order of any of its subgroups. I state each such theorem in the text the first time it is used. Often an alternative, even more algebraic approach to a topic covered in the text is treated in the problems.

Appendix 1 contains some essential algebra, which is not all rudimentary and may be new to the reader: the traditional definitions of the theory of divisibility in an integral domain and a proof that a Euclidean domain enjoys unique factorization. That theorem is applied to the integers in Chapter 1, to the ring of polynomials with coefficients in a field in Chapter 3, and to some rings of algebraic integers defined in Chapter 6.

We have all been familiar with the arithmetic of the integers since elementary school, so the study of number theory is an ideal place to discover that mathematics is an experimental science. The subject of our experiments is the well-known sequence $1, 2, 3, \ldots$; the results of those experiments are theorems which show that observed patterns and regularities are not coincidental. This book, like most, almost always gives only the theorems and suppresses the experimental evidence that would be costly to include and dull to read. The reader is urged to reconstruct it by computing numerical special cases of each definition and theorem. To encourage this habit many of the problems begin "Investigate ..." rather than "Prove. ..."

The problems are important and often difficult. They consist of applications and examples of theorems and techniques in the text, numerical examples which show that arguments have been pushed to their natural boundaries, special cases of topics treated later in the book, and material often included in more traditional books on elementary number theory. The more time the reader spends on them the better. I have marked the harder or longer problems with an asterisk, but that subjective classification is not always reliable. A starred problem may yield to a special trick while one I think easy proves surprisingly stubborn.

Sections are numbered consecutively through the book, equations consecutively within each chapter. The notation $m.n$ refers to the nth numbered item in Section m; it may be a theorem, definition, lemma, corollary, or example. The bibliography lists only the general works on which I relied most heavily; other references occur where relevant in the text and problems.

Finally, I should like to thank my algebra class at Bryn Mawr, which suffered through false starts while I learned the number theory I was teaching; Mary Wolfe, whose lecture notes were an invaluable zeroth draft of the manuscript; Bryn Mawr College, for generous support during a leave of absence; William Adams, who read and commented on an early version of the manuscript; Russ Fallowes, who wrote the program that produced Appendix 3; and my wife Joan for many kinds of aid and comfort, from the routine of reading proof to the sublime.

Ethan D. Bolker

Bryn Mawr, Pennsylvania
April 1969

To my parents

Contents

1

Linear Diophantine Equations

We shall begin our study of number theory not with the topic announced in the title of this first chapter but with an empirical investigation of a problem studied and solved by Fermat.

1. SUMS OF SQUARES

Which positive integers can be written as sums of two integral squares? Let us call such integers *representable*; the first few representable integers are 1, 2, 4, 5, 8, 9, 10, 13, and 16. What pattern does the sequence of representable integers form? How can we decide whether a given integer is representable? We need more data to make intelligent guesses.

The representable integers less than 100 appear in boldface in Table 1. The arrangement of that table in rows of eight allows us to guess a part of the pattern; the third, sixth, and seventh columns seem free of representable integers. The reader is invited (in Problem 6.1) to prove that that phenomenon continues. It is harder to predict which integers in the other columns are representable. Some hints will be found in Problems 6.2 and 6.3.

Some integers can be represented more than one way. The smallest such is $25 = 5^2 + 0^2 = 4^2 + 3^2$; the others less than 100 are 50, 65, and 85.

TABLE 1ᵃ

1	2	3	4	5	6	7	8
1	**2**	3	**4**	**5**	6	7	**8**
9	**10**	11	**12**	**13**	14	15	**16**
17	**18**	19	**20**	21	22	23	24
25	**26**	27	28	**29**	30	31	**32**
33	**34**	35	**36**	37	38	39	**40**
41	42	43	44	**45**	46	47	48
49	**50**	51	**52**	**53**	54	55	56
57	**58**	59	60	**61**	62	63	**64**
65	66	67	**68**	69	70	71	**72**
73	**74**	75	76	77	78	79	**80**
81	**82**	83	84	**85**	86	87	88
89	**90**	91	92	93	94	95	96
97	**98**	99	**100**				

ᵃ Integers in boldface type are representable as sums of two squares.

The least integer representable three ways is

$$325 = 18^2 + 1^2$$
$$= 17^2 + 6^2$$
$$= 15^2 + 10^2.$$

Counting the number of ways n can be represented gives clues to the pattern of representable integers.   (See Problems 6.2 and 6.3.)

The recorded history of the study of representable integers starts in about 250 A.D. with Diophantos.   Problems in which integral values of the unknowns are sought are called *Diophantine* in his honor.   In the seventeenth century Fermat gave, as a simple function of n, the number of solutions to the Diophantine equation $x^2 + y^2 = n$ and thus answered all the questions we have raised about the representability of integers.   His answer is our Theorem 36.3.

In the preface to the collected works of Eisenstein, Gauss wrote of number theory that "... important propositions, with the impress of simplicity upon them, are often easily discoverable by induction and yet are of so profound a character that we cannot find their demonstration until after many vain attempts; and even then, when we do suceed, it is often by some tedious and artificial process, while the simpler methods may long remain concealed." The argument which stretches from here to our proof of Fermat's theorem on representable integers and beyond is not the shortest possible, but by lengthening and modernizing it we have freed it of many of the "artificial processes" and revealed the "simpler methods" to which Gauss refers.

2. DIVISIBILITY AND UNIQUE FACTORIZATION

The simplest nontrivial Diophantine equation,

$$ax = b, \ (a \neq 0) \tag{1}$$

has a solution if and only if a *divides* b; when that occurs we write $a \mid b$ and say too that a is a *factor* or *divisor* of b, while b is a *multiple* of a. Since $a0 = 0$, we have $a \mid 0$ for every a. An integer p other than 0, 1, or -1 is *prime* when its only divisors are ± 1 and $\pm p$. If $n \neq 0$, ± 1 is not prime, it is composite.

2.1 Theorem. Every nonzero integer other than ± 1 can be written "uniquely" as a product of primes.

The uniqueness is qualified by the quotation marks because, for example,

$$6 = (-2)(-3) = 3 \cdot 2$$

and 2, -2, 3, and -3 are all primes. What we mean by "uniquely" is, of course, that if $n = \pm p_1 \cdots p_r = \pm q_1 \cdots q_s$ where the p_i and the q_j are primes, then $r = s$, and the sequence $|p_1|, \ldots, |p_r|$ is a permutation of the sequence $|q_1|, \ldots, |q_s|$.

In Appendix 1 we prove that Lemma 2.2 below implies Theorem 2.1, the fundamental theorem of arithmetic. We repeat here some of the auxiliary definitions and intermediate steps in that argument.

2.2 Lemma. Given integers m and $n \neq 0$ there are unique integers q and r with $0 \leq r < |n|$ for which

$$m = qn + r.$$

This is simply a formal assertion of the possibility of "division with remainder" in **Z**, the ring of integers.

2.3 Definition. The integer d is a *greatest common divisor* of a and b in **Z** if $d \mid a$ and $d \mid b$, and if whenever $c \mid a$ and $c \mid b$, then $c \mid d$.

2.4 Theorem. Every pair $\langle a, b \rangle$ of integers except $\langle 0, 0 \rangle$ has a greatest common divisor d. The Diophantine equation

$$ax + by = d \tag{2}$$

has a solution.

There are several algorithms for computing d and solving Eq. (2) in a predictable finite number of steps. The most common, the *Euclidean algorithm*, is illustrated in Example 15 of Appendix 1.

If d is a greatest common divisor of a and b, then so is $-d$; no other integer enjoys this privilege. We shall write (a, b) for the positive greatest common divisor of a and b and $\langle a, b \rangle$ for the ordered pair formed by a and b. Thus $(4, 6) = 2$ and $(25, -4) = 1$. If $a \neq 0$ then $(a, 0) = |a|$ because the common divisors of a and 0 are just the divisors of a; $(0, 0)$ is undefined.

2.5 Definition. The integers a and b are *relatively prime*, or *coprime*, if and only if $(a, b) = 1$.

If p is prime and $a \neq \pm 1$, then p and a are relatively prime if and only if $p \nmid a$, that is, p does not divide a.

2.6 Theorem. If $a \mid bc$ and $(a, b) = 1$, then $a \mid c$. In particular, if a prime divides a product, then it divides one of the factors. (See Appendix 1, Theorem 16.)

2.7 Definition. An integer $m \neq 0$ is a *common multiple* of a and b if $a \mid m$ and $b \mid m$. If neither a nor b is 0, then $|ab|$ is a common multiple, so among the positive common multiples there is a least, which we write

$$\text{l.c.m.} \{a, b\}.$$

2.8 Theorem. If $ab > 0$, then the least common multiple of a and b is $ab/(a, b)$; the least common multiple divides any common multiple.

3. THE DIOPHANTINE EQUATION $ax + by = c$

The second simplest Diophantine equation is $ax + by = c$ which one often meets in secondary school in a form such as: "How many dimes and quarters make n cents?" That banking problem is clearly hopeless unless $5 \mid n$.

3.1 Theorem. The Diophantine equation

$$ax + by = c, \qquad \langle a, b \rangle \neq \langle 0, 0 \rangle \tag{3}$$

has solutions if and only if $(a, b) \mid c$. When solutions exist, they are all given by

$$x = \frac{c}{(a, b)} x_0 + \frac{b}{(a, b)} n$$

$$y = \frac{c}{(a, b)} y_0 - \frac{a}{(a, b)} n \tag{4}$$

where

$$ax_0 + by_0 = (a, b) \tag{5}$$

and n is any integer.

Proof. "Only if" is obvious. To show that Eqs. (4) give all the solutions suppose $(a, b) \mid c$ and that $\langle x, y \rangle$ is a solution to Eq. (3). Let $a' = a/(a, b)$; $b' = b/(a, b)$; and $c' = c/(a, b)$. Use Theorem 2.4 to find a solution $\langle x_0, y_0 \rangle$ to Eq. (5). Then

$$a'(x - c'x_0) = -b'(y - c'y_0). \tag{6}$$

Since $(a', b') = 1$ (Problem 6.11), Eq. (6) implies

$$a' \mid y - cy_0$$

so that

$$y = c'y_0 + a'n$$

for some integer n. Then

$$a'(x - c'x_0) = -a'b'n$$

so

$$x = c'x_0 - b'n. \tag{7}$$

Moreover, if x and y are defined by Eqs. (4) then $\langle x, y \rangle$ solves Eq. (3).

4. THE DIOPHANTINE EQUATION $a_1 x_1 + \cdots + a_n x_n = c$

Let $a_1, \ldots, a_n$ be integers at least one of which is not 0. A greatest common divisor d of $a_1, \ldots, a_n$ is a common divisor which is a multiple of every common divisor. Write $(a_1, \ldots, a_n)$ for the unique positive greatest common divisor the proof of whose existence is Problem 6.13. Now we generalize Theorem 3.1.

4.1 Theorem. The Diophantine equation

$$a_1 x_1 + \cdots + a_n x_n = c \tag{8}$$

(some $a_i \neq 0$) has a solution if and only if $(a_1, \ldots, a_n) \mid c$.

Proof. When $n = 2$, this is just Theorem 3.1, so our induction on n, the number of variables, begins well. Suppose $n > 2$. We shall show that Eq. (8) can be solved assuming that similar equations in $n - 1$ variables can be solved. To this end observe that $(a_1, a_2, \ldots, a_n) = ((a_1, a_2), a_3, \ldots, a_n)$ (Problem 6.14). Then use Theorem 3.1 and the inductive hypothesis to find integers $y_1, y_2, x, x_3, \ldots, x_n$ satisfying

$$a_1 y_1 + a_2 y_2 = (a_1, a_2)$$

and

$$(a_1, a_2)x + a_3 x_3 + \cdots + a_n x_n = c.$$

If we set $x_1 = y_1 x$ and $x_2 = y_2 x$, then $\langle x_1, \ldots, x_n \rangle$ solves Eq. (8).

5. THE INFINITUDE OF THE PRIMES

This section contains some information on the way in which the primes are distributed in **Z** and introduces some techniques which we shall generalize in the next chapter. Euclid knew the following theorem.

5.1 Theorem. There are infinitely many primes.

Proof. Let P be the set of primes. Since 2 is prime, P is not empty. We shall show that no finite subset Q of P exhausts P. Suppose $Q = \{q_1, \ldots, q_n\}$ is a nonempty subset of P. Let $m = 1 + q_1 \cdots q_n \neq \pm 1$. The fundamental theorem of arithmetic implies that there is a prime p which divides m. Since no q_i divides m, $p \notin Q$. That is, $Q \neq P$, so P is infinite.

We can exploit this method to prove a little more. Lemma 2.2 implies that every integer n may be written in just one of the forms, $4k$; $4k + 1$; $4k + 2$; or $4k + 3$. Since $2 \mid 4k$ and $2 \mid 4k + 2$, every odd prime is of the form $4k + 1$ or $4k + 3$. The following lemma is a very special case of a principle we shall introduce in Section 7.

5.2 Lemma. The product of integers of the form $4k + 1$ is again of that form.

Proof. It clearly suffices to prove this lemma for the product of just two integers. Let $n = 4k + 1$ and $n' = 4k' + 1$. Then

$$
\begin{aligned}
nn' &= (4k + 1)(4k' + 1) \\
&= 16kk' + 4k + 4k' + 1 \\
&= 4(4kk' + k + k') + 1.
\end{aligned}
$$

5.3   Theorem.   There are infinitely many primes of the form $4k + 3$.

Proof.   Let P be the set of such primes.   Since $3 \in P$, P is not empty.   We shall show that no finite subset Q of P exhausts P.   Suppose $Q = \{q_1, \ldots, q_n\}$ is a nonempty subset of P.   Let

$$m = 4q_1 \cdots q_n - 1$$
$$= 4(q_1 \cdots q_n - 1) + 3 \neq \pm 1.$$

The fundamental theorem of arithmetic implies that m is a product of primes. Since m is odd, every prime divisor of m is odd.   If all the prime divisors of m were of the form $4k + 1$, then m would be of that form (Lemma 3.2), which is false.   Therefore there is a prime $p \in P$ which divides m.   Since no q_i divides m, $p \notin Q$.   That is $Q \neq P$; P is infinite.

Note that we have not used the full strength of the fundamental theorem of arithmetic.   We needed only the existence, not the "uniqueness," of a prime factorization of m to prove Theorems 5.1 and 5.3.

The technique we used to prove Theorem 5.3 fails to prove the infinitude of the primes of the form $4n + 1$ since the analogue of Lemma 3.2 is false: $(4 \cdot 1 + 3)(4 \cdot 0 + 3) = 7 \cdot 3 = 21 = 4 \cdot 5 + 1$.   In fact, the product of two numbers of the form $4k + 3$ is always of the form $4k + 1$.   There *are* infinitely many primes of the form $4k + 1$; we shall prove it in Section 12.   Much more is known.   A justly famous theorem of Dirichlet's asserts that the arithmetic progression of numbers of the form $ak + b$ contains infinitely many primes whenever $a \neq 0$ and $(a, b) = 1$.   Dirichlet's theorem is beyond the scope of this book; we shall, however, prove several special cases.

6. PROBLEMS

6.1   Show that the third, sixth, and seventh columns of Table 1 remain free of representable integers.

6.2   Prove that $2n$ is representable when n is. Is the converse true?

6.3   Prove that mn is representable when m and n are.   How many ways can mn be represented?

6.4*   Formulate some guesses about integers which can be represented as a sum of three squares.   Show that column 7 in Table 1 contains no such integers.   We shall show in Section 40 that every integer is a sum of four squares.

6.5   Show $n(n - 1)(2n - 1)$ is always divisible by 6.

6.6   Prove that the Diophantine equation

$$3y^2 - 1 = x^2$$

has no solutions.   That is, $3y^2 - 1$ is never a perfect square.

6.7   Prove: If $2 \nmid n$ and $3 \nmid n$, then $24 \mid (n^2 + 23)$.

6.8   What is $(3^2 \cdot 5^6 \cdot 7^3 \cdot 17,\ 3^{21} \cdot 5 \cdot 13)$?   The answer is easily computed without the Euclidean algorithm and suggests a useful theorem for computing greatest common divisors when prime factorizations are known.   That theorem simplifies some of the problems which follow.

6.9   Prove Theorem 2.8.

6.10   Prove: If $(a, b) = 1$ and ab is a square, then $|a|$ and $|b|$ are squares.

6.11   Prove $\left(\dfrac{a}{(a, b)},\ \dfrac{b}{(a, b)} \right) = 1$.

6.12   Use the fundamental theorem of arithmetic to show that the Diophantine equation $nx^2 = y^2$ has a solution if and only if n is a square.   Deduce that $\sqrt{n}$ is irrational unless n is a square.

6.13   Prove that n integers at least one of which is not zero have a greatest common divisor.   (Either apply Theorem 2.4 inductively or, better, mimic the proof of Theorem 14 in Appendix 1.)

6.14   Prove $((a_1, a_2),\ a_3, \ldots, a_n) = (a_1, a_2, \ldots, a_n)$ when $a_1 \neq 0$.

6.15   Solve the Diophantine equations

$$153x - 34y = 51$$

and

$$30x + 105y + 70z + 42w = 1.$$

6.16*   Given integers a, b, and c discuss the existence of solutions $\langle x, y \rangle$ to $ax + by = c$ for which x and y are positive.   Some elementary analytic geometry may help.

6.17   Let $F_n = 2^n + 1$.   If F_n is prime, it is called a *Fermat prime*.   Prove: If F_n is prime, then n is a power of 2. F_1, F_2, F_4, F_8, and $F_{16} = 65{,}537$ are prime. No other Fermat primes are known.   Euler showed $641 \mid F_{32}$.

6.18   Let $M_n = 2^n - 1$.   If M_n is prime, it is called a *Mersenne prime*.   Prove: If M_n is prime, then n is prime. M_{11213} is the largest known prime.   We do not know whether there are infinitely many Mersenne primes (D. B. Gillies, "Three New Mersenne Primes and a Statistical Theory," *Mathematics of Computation*, **18** (1964) 93–97).

6.19   Show that there are infinitely many primes of the form $6k + 5$.

6.20   Show that for every $n > 2$ there are infinitely many primes which are *not* of the form $nk + 1$, so that, for example, the set $\{8k + v \mid k \in \mathbf{Z},\ v = 3,\ 5,\ 7\}$ contains infinitely many primes.

6.21   Suppose $(m, n) = 1$ so that the fraction m/n is written in "lowest terms."

Show that

$$\frac{m}{n} = x^2 + y^2$$

has a solution in rational numbers x and y if and only if n is a square and m is representable.

6.22 Find 15 consecutive composite integers. Find n consecutive composite integers.

2

Congruence

Gauss first formalized the concept of congruence we are about to introduce. That concept now pervades number theory, for it provides an elegant and efficient tool for studying and solving Diophantine problems.

7. ARITHMETIC IN Z_n. SOLVING CONGRUENCES

We say integers a and b are *congruent modulo n* when $n \mid a - b$; we then write $a \equiv b(n)$. Thus, for example,

$$3 \equiv 7 \equiv -1(4)$$

$$5 \not\equiv 6(2).$$

Note that $a \equiv b(-n)$ if and only if $a \equiv b(n)$ so that we may restrict our attention to congruence modulo positive integers n.

The language of congruence is a good one in which to express the results at the end of Section 5, since p is of the form $4k + 3$ if and only if $p \equiv 3(4)$. Thus Theorem 5.3 reads: "There are infinitely many primes congruent to 3 modulo 4."

The set $n\mathbf{Z}$ of multiples of n in $\mathbf{Z}$ consists precisely of the integers congruent to 0 modulo n. The set $n\mathbf{Z}$ is an ideal in $\mathbf{Z}$ and $a \equiv b(n)$ if and only if

$a - b \in n\mathbf{Z}$, that is, if and only if a and b are in the same coset of $n\mathbf{Z}$. Thus congruence is an equivalent relation, and a congruence class is just a coset of $n\mathbf{Z}$ in $\mathbf{Z}$. Each such coset contains precisely all the integers in an arithmetic progression of period n. This connection between congruence modulo n and the cosets of $n\mathbf{Z}$ is often used to motivate the theory of abstract groups and rings. We shall now use that theory to shed light on this motivating example. We write $\mathbf{Z}_n$ for the quotient ring $\mathbf{Z}/n\mathbf{Z}$ and $a \rightsquigarrow \bar{a}$ for the natural ring homomorphism of $\mathbf{Z}$ onto $\mathbf{Z}_n$. That is, $\bar{a}$ is the coset $n\mathbf{Z} + a$ of $n\mathbf{Z}$ to which a belongs. The zero element of $\mathbf{Z}_n$ is $n\mathbf{Z} = \bar{0}$, and $a \equiv b(n)$ if and only if $\bar{a} = \bar{b}$.

Suppose m is an integer. The division with remainder Lemma (2.2) implies the existence of unique integers q and r such that $0 \leq r < n$ and $m = qn + r$. Then $m \equiv r(n)$. Moreover, no two integers between 0 and $n - 1$ inclusive can be congruent modulo n. Thus the set $\{0, 1, \ldots, n - 1\}$ is a complete set of representatives of the cosets of $n\mathbf{Z}$ in $\mathbf{Z}$; $\mathbf{Z}_n$ has n elements. We shall often refer to these elements as $0, 1, \ldots, n - 1$ instead of $\bar{0}, \ldots, \overline{n - 1}$. We shall, of course, avoid this convention when confusion might result.

Much of this chapter and Chapter 4 is devoted to the study of the rings $\mathbf{Z}_n$. The definition of the arithmetic in those quotient rings depends upon the fact that if $a \equiv a'(n)$ and $b \equiv b'(n)$, then $a + b \equiv a' + b'(n)$ and $ab \equiv a'b'(n)$. These assertions are true because $n\mathbf{Z}$ is an ideal in $\mathbf{Z}$. The reader should take the time now to verify them by direct computation. Note that these remarks generalize Lemma 5.2.

The additive group $\mathbf{Z}_n$, $+$ of the ring $\mathbf{Z}_n$ is cyclic since 1, or, rather, $\bar{1}$, is a generator. We shall have more to say about cyclic groups in Section 17. The multiplicative structure of $\mathbf{Z}_n$ is more complicated and more interesting. Consider the problem: Find an integer x such that

$$ax \equiv b\ (n). \tag{1}$$

We can usefully reformulate this problem two ways. First it is really a problem in $\mathbf{Z}_n$, rather than in $\mathbf{Z}$, for it asks for a solution in $\mathbf{Z}_n$ to the equation

$$AX = B \tag{2}$$

where $A = \bar{a}$ and $B = \bar{b}$. If X solves Eq. (2), then any $x \in X$ solves Eq. (1). Conversely, if x solves Eq. (1), then $\bar{x}$ solves Eq. (2).

Second, Eq. (1) is a Diophantine equation, for it has a solution if and only if there are integers x and y such that

$$ax = b + ny. \tag{3}$$

We shall often pass casually from one to another of these three formulations of a problem: solve a congruence, solve an equation in $\mathbf{Z}_n$, and solve a Diophantine equation.

In the particular problem we set ourselves in Eq. (1) the third formulation is the most congenial, for we have already solved it! Theorem 3.1 says that Eq. (3) has a solution if and only if

$$(a, n) \mid b$$

and that then any two solutions differ by a multiple of $n/(a, n)$. Thus the solutions to Eq. (1) form one coset of $(n/(a, n))\mathbf{Z}$. These remarks prove the following theorem.

7.1 Theorem. $ax \equiv b(n)$ has a solution if and only if $(a, n) \mid b$. When solutions exist they are unique modulo $n/(a, n)$.

For example, to solve

$$6x \equiv 4(10)$$

solve

$$6x + 10y = 4$$

instead. This is equivalent to

$$3x + 5y = 2.$$

Now

$$3 \cdot 2 + 5 \cdot (-1) = 1$$

(by inspection or by the Euclidean algorithm illustrated in Appendix 1). So

$$3 \cdot 4 + 5 \cdot (-2) = 2$$

and $x = 4$ solves the original congruence. The other solutions are the integers $4 + (10/(6, 10))k = 4 + 5k$, so that $\ldots, -6, -1, 4, 9, 14, \ldots$ are all the solutions.

Note that these solutions fill just one coset of $5\mathbf{Z}$ but two cosets of $10\mathbf{Z}$. That is, our original congruence has the two distinct solutions 4 and 9 (or, formally, $\bar{4}$ and $\bar{9}$) when it is reformulated as a problem in $\mathbf{Z}_{10}$. This phenomenon is examined in greater generality and abstraction in Problem 11.5.

8. THE CHINESE REMAINDER THEOREM

In this section we shall use Theorem 4.1 to study simultaneous congruences in one unknown.

8.1 Chinese Remainder Theorem. Let $n_1, \ldots, n_r, b_1, \ldots, b_r$ be integers such that $(n_i, n_j) = 1$ when $i \neq j$. Then the congruences

$$x \equiv b_i (n_i) \quad (i = 1, \ldots, r) \tag{4}$$

have a simultaneous solution.

Before proving the Chinese remainder theorem we shall discuss it briefly. Some hypotheses about the moduli n_i are necessary lest the congruences in Eq. (4) interfere with one another. For example, the two congruences

$$x \equiv 1\,(2) \quad \text{and} \quad x \equiv 2\,(4)$$

cannot have a simultaneous solution since only odd integers solve the first and only even integers the second. However, the condition that the n_i be mutually relatively prime is not necessary. For example, the congruences $x \equiv 1(n_i)$ have the simultaneous solution $x = 1$ for any choice of $n_1, \ldots, n_r$. Two stronger versions of the Chinese remainder theorem can be found in Problems 11.7 and 11.8.

The solutions to $x \equiv b_i(n_i)$ form an arithmetic progression with period n_i. Thus the Chinese remainder theorem is equivalent to the statement that r arithmetic progressions with pairwise relatively prime periods have a non-empty intersection. This is a theorem more appropriately stated in terms of congruences rather than in terms of the arithmetic of $\mathbf{Z}_n$ since it concerns congruences with respect to different moduli. It describes how the cosets of the ideals $n_i\mathbf{Z}$ fit together in $\mathbf{Z}$.

Proof (of the Chinese remainder theorem). Suppose we could find integers $x_1; \ldots, x_r$ for which $x_i \equiv 1(n_i)$ but $x_i \equiv 0(n_j)$ when $1 \leq i \neq j \leq r$. Then $x = b_1 x_1 + \cdots + b_r x_r$ would satisfy all the congruences in (4). To find such x_i let $N = n_1 \cdots n_r$ and $N_i = N/n_i$. Then $(N_1, N_2) = N/n_1 n_2$ since $(n_1, n_2) = 1$. Continuing inductively, then $(N_1, \ldots, N_r) = 1$, so we can use Theorem 4.1 to find integers $y_1, \ldots, y_r$ such that

$$N_1 y_1 + \cdots + N_r y_r = 1. \tag{5}$$

Finally, let $x_i = N_i y_i$. Since $i \neq j$ implies $n_i \mid N_j$, $x_j \equiv 0(n_i)$. Now reduce Eq. (5) modulo n_i. All the terms but $\overline{x_i}$ are zero, so $x_i \equiv 1(n_i)$.

8.2 Example. Later in the book we shall use the special case of Theorem 8.1 in which $b_2 = b_3 = \cdots = b_r = 1$. That special case is easily treated by a special trick. Let $N_1 = n_2 \cdots n_r$. Then for every integer k and $2 \le i \le r$, $x = kN_1 + 1 \equiv 1(n_i)$. Thus we need only adjust k so that

$$x - 1 = kN_1 \equiv b_1 - 1(n_1).$$

Such a k exists because $(n_1, N_1) = 1$ (Theorem 7.1). For example, let us solve the simultaneous congruences

$$x \equiv 2 \ (9) \qquad x \equiv 1 \ (5) \qquad x \equiv 1 \ (7).$$

Here $N_1 = 5 \cdot 7 = 35$ and $x = 35k + 1$. We wish to choose k so that

$$35k \equiv 2 - 1 \ (9)$$

or equivalently

$$-k \equiv 1 \ (9).$$

Clearly $k = -1$ will do. Thus $x = -34$ solves our problem. If a positive answer is preferred, let

$$x = -34 + 9 \cdot 5 \cdot 7 = -34 + 315 = 281.$$

8.3 Example. The Chinese remainder theorem enables us to reduce questions about congruences modulo a composite integer n to questions about congruences modulo the prime power divisors of n. Suppose we wish to solve

$$x^2 \equiv -1 \ (325). \tag{6}$$

Since $325 = 5^2 \cdot 13$ we consider instead

$$x^2 \equiv -1 \ (25)$$
$$x^2 \equiv -1 \ (13). \tag{7}$$

By inspection we can see that 7 solves the first of these and 5 the second. Now let us use the Chinese remainder theorem to solve

$$x \equiv 7 \ (25)$$
$$x \equiv 5 \ (13) \tag{8}$$

simultaneously. Here $n_1 = 25 = N_2$, and $n_2 = 13 = N_1$. Since

$$2 \cdot 13 + (-1) \cdot 25 = 1$$

let

$$x = 7 \cdot 2 \cdot 13 + 5(-1)25 = 57.$$

Then 57 satisfies both congruences in (8), hence both in (7), hence (6).

9. THE EULER φ-FUNCTION

In this section we begin the study of the multiplicative structure of Z_n which will continue through this and the next two chapters. Let R be a commutative ring with a (multiplicative) identity $1 \neq 0$. R^* is the set of nonzero elements of R; R is an *integral domain* if and only if R^* is closed under multiplication. The ring R is a *field* if and only if R^* is a multiplicative group, and $u \in R$ is a *unit* if and only if it has a multiplicative inverse; the set of units is contained in R^* and is a group under multiplication (see Appendix 1, Lemma 3).

9.1 Definition. When $n > 1$, $\Phi(n)$ is the group of units of Z_n; $\varphi(n)$ is the order of $\Phi(n)$. Set $\varphi(1) = 1$.

The function we have just introduced is the famous φ-function introduced by Euler. We shall soon know a great deal more about it.

We have already built the machinery which allows us to recognize the units in Z_n.

9.2 Theorem. $\bar{a} \in \Phi(n)$ if and only if $(a, n) = 1$.

Proof. $\bar{a} \in \Phi(n)$ if and only if the equation $\bar{a}X = 1$ has a solution in Z_n or, equivalently, $ax \equiv 1(n)$ has a solution in Z. But this happens if and only if $(a, n) = 1$ (Theorem 7.1). It follows that Definition 9.1 is equivalent to Euler's, for Theorem 9.2 together with the identification of Z_n with $\{1, \ldots, n - 1\}$ yields the following corollary.

9.3 Corollary. If $n > 1$, then $\varphi(n)$ is the number of positive integers less than n and relatively prime to n.

For example, 1, 5, 7, and 11 are the units in Z_{12} and $\varphi(12) = 4$. Strictly speaking, $\bar{1}, \bar{5}, \bar{7}$ and $\overline{11}$ are the units; this is the last time we shall point out the distinction unless it is necessary to avoid confusion.

9.4   Corollary. p is prime if and only if $\varphi(p) = p - 1$.

Proof. p has no proper factors if and only if none of the integers $1, 2, \ldots, p - 1$ shares a factor with p.

9.5   Theorem. p is prime if and only if $\mathbf{Z}_p$ is a field.

Proof. $\mathbf{Z}_p$ is a field if and only if each of the $p - 1$ elements of $\mathbf{Z}_p^*$ is a unit, that is, if and only if $\varphi(p) = p - 1$.

Here is a second proof of half the theorem. Suppose p is prime. If $(a, p) = (b, p) = 1$, then Theorem 2.6 implies $(ab, p) = 1$. Consequently $\mathbf{Z}_p^*$ is closed under multiplication, so $\mathbf{Z}_p$ is an integral domain. But every finite integral domain is a field (Problem 11.11).

We can now deduce some classical parts of elementary number theory from familiar facts about groups. Let G be a group with identity element e and $g \in G$. Then the order of g divides the order n of G, so that $g^n = e$ and $g^{n-1} = g^{-1}$.

9.6   Euler's Theorem. $k^{\varphi(n)} \equiv 1(n)$ when $(k, n) = 1$.

Proof. $\varphi(n)$ is the order of the group $\Phi(n)$ which contains k. Thus $k^{\varphi(n)}$ is the identity element of $\Phi(n)$.

9.7   Fermat's (Little) Theorem. If p is prime then $k^{p-1} \equiv 1(p)$ when $p \nmid k$.

Proof. Let p be prime. Then $\varphi(p) = p - 1$ and $(p, k) = 1$ if $p \nmid k$. Thus Fermat's theorem is a special case of Euler's theorem.

9.8   Fermat's (Little) Theorem, second version. If p is prime, then $k^p \equiv k(p)$.

Proof. If $p \mid k$, then $p \mid k^p$, so $k^p \equiv 0 \equiv k(p)$. If $p \nmid k$, then $k^{p-1} \equiv 1(p)$, so multiplying through by k,

$$k^p \equiv k \ (p).$$

Fermat's (great) or (last) "Theorem" is the subject of Chapter 7.

We can sometimes use Euler's theorem to shorten the solutions of congruences. If $(a, n) = 1$, then $ax \equiv b(n)$ has a solution. In $\mathbf{Z}_n$

$$\bar{a}\bar{x} = \bar{b}$$

so

$$\bar{x} = \bar{b}\bar{a}^{-1}$$
$$= \bar{b}\bar{a}^{\varphi(n)-1} \qquad \text{(Euler's theorem)}$$
$$= \overline{ba^{\varphi(n)-1}}.$$

Thus $x = ba^{\varphi(n)-1}$ is a solution to the original congruence. For example,

$$5x \equiv 4 \,(12)$$

has

$$x = 4 \cdot 5^{\varphi(12)-1} = 4 \cdot 5^3 = 500$$

as a solution.   Of course, since

$$500 = 41 \cdot 12 + 8$$
$$500 \equiv 8 \equiv -4 \,(12)$$

so $x = -4$ and $x = 8$ are also solutions.

We can also prove the Chinese remainder theorem (8.1) in the following new way.   Suppose $(n_i, n_j) = 1$ when $i \ne j$.   As in the proof in Section 8, let $N = n_1 \cdots n_r$ and $N_i = N/n_i$.   Then $(n_i, N_i) = 1$ and $N_i \equiv 0(n_j)$ when $i \ne j$.   Let

$$x = b_1 N_1^{\varphi(n_1)} + \cdots + b_r N_r^{\varphi(n_r)}.$$

Then

$$x \equiv b_i N_i^{\varphi(n_i)} \equiv b_i \,(n_i)$$

and Theorem 8.1 is proved.   This construction of x has the computational advantage of requiring no long division, though x is likely to be large.

10.   MORE ABOUT $\varphi(n)$

Consider the n rational numbers

$$\frac{1}{n}, \frac{2}{n}, \ldots, \frac{n}{n}. \tag{9}$$

How many of these fractions are written in lowest terms? Since m/n is in lowest terms if and only if $(m, n) = 1$, the answer to this question is $\varphi(n)$. Suppose now that we reduce each fraction in (9) to lowest terms. The denominators which occur when we have finished are just the divisors of n. How often does each divisor appear? If $d \mid n$, then each of the rational numbers $1/d, \ldots, d/d$ is m/n for some m and hence appears in (9). However, of these d fractions with denominator d just $\varphi(d)$ are in lowest terms. Therefore the denominator d appears just $\varphi(d)$ times when the fractions in (9) are written in lowest terms. But all the fractions have now been accounted for, so n, the number of fractions, is just the sum of the numbers $\varphi(d)$ for d dividing n. We have notation summarizing that clumsy sentence:

$$n = \sum_{d \mid n} \phi(d). \tag{10}$$

Note that implicit in Eq. (10) are the assumptions that n and d are positive and that $d = 1$ and $d = n$ are counted as divisors.

If we rewrite Eq. (10) as

$$\varphi(n) = n - \sum_{\substack{d \mid n \\ d < n}} \varphi(d) \tag{11}$$

it becomes a formula for computing $\varphi(n)$ recursively in terms of $\varphi(d)$ for the proper divisors d of n.

The reader should work through the derivation of Eq. (10) for some particular values of n. The value $n = 12$ is good because 12 is the smallest number with a varied set of divisors.

For primes p, Eq. (11) becomes

$$\varphi(p) = p - \varphi(1) = p - 1, \tag{12}$$

an old friend. All the fractions in the sequence (9) except p/p are in lowest terms.

We can use Eq. (11) to gain new knowledge. If p is prime, then the divisors of p^2 are just 1, p, and p^2, so that

$$\begin{aligned}
\varphi(p^2) &= p^2 - \varphi(p) - \varphi(1) \\
&= p^2 - (p - 1) - 1 \\
&= p^2 - p.
\end{aligned}$$

We employ this device inductively in the next theorem.

10.1   Theorem.   If p is prime and $\alpha \geq 1$, then

$$\varphi(p^\alpha) = p^\alpha - p^{\alpha-1}.$$

Proof.   We already know this theorem is true for $\alpha = 1$ and $\alpha = 2$.   Suppose that it is true for all $\alpha < \beta$.   The proper divisors of p^β are just $1, p, \ldots, p^{\beta-2}$, and $p^{\beta-1}$.   Therefore, Eq. (11) implies

$$\begin{aligned}
\varphi(p^\beta) &= p^\beta - [\varphi(p^{\beta-1}) + \varphi(p^{\beta-2}) + \cdots + \varphi(p) + 1] \\
&= p^\beta - [(p^{\beta-1} - p^{\beta-2}) + (p^{\beta-2} - p^{\beta-3}) + \cdots + (p - 1) + 1] \\
&\quad \text{(the inductive hypothesis)} \\
&= p^\beta - p^{\beta-1}.
\end{aligned}$$

Proof (Second version).   The telescoping sum above is pretty but unenlightening.   We shall give another proof of Theorem 10.1 which is a straightforward counting argument.   A positive integer $k < p^\beta$ satisfies $(k, p^\beta) > 1$ if and only if k is a multiple of p.   How many multiples of p less than p^β are there?

$$1 \leq mp < p^\beta \Leftrightarrow 1 \leq m < p^{\beta-1}$$

so that of the p^β integers $1, 2, \ldots, p^\beta - 1$ exactly $p^{\beta-1} - 1$ share a factor with p^β.   The others are in $\Phi(p^\beta)$.   Therefore

$$\varphi(p^\beta) = (p^\beta - 1) - (p^{\beta-1} - 1) = p^\beta - p^{\beta-1}.$$

Notice that we used Greek letters for exponents in Theorem 10.1.   We shall try to adhere to this convention in the following pages.

10.2   Definition.   A function $\Psi : \mathbf{Z} \to \mathbf{Z}$ is *multiplicative* if and only if $\Psi(mn) = \Psi(m)\Psi(n)$ whenever m and n are relatively prime.

The identity function is multiplicative; so is the absolute value function, since for these Ψ, $\Psi(mn) = \Psi(m)\Psi(n)$ for all m and n.   Problems 11.18, 11.25, and 11.29 consider examples of multiplicative functions.   Problem 11.30 concerns the theory of such functions.

A multiplicative function Ψ is known when its values for prime powers are known, because if $n = p_1^{\alpha_1} \cdots p_k^{\alpha_k}$, then

$$\Psi(n) = \Psi(p_1^{\alpha_1}) \cdots \Psi(p_k^{\alpha_k}).$$

Our next task is to show φ is multiplicative.

10.3 Lemma.   Suppose $(m, n) = 1$.   Then

$$(xm + yn, mn) = (x, n)(y, m). \tag{13}$$

Proof.   It suffices to show that the two members of Eq. 13 have the same prime power divisors.   Suppose p^α divides either member.   Then $p^\alpha \mid mn$. Since $(m, n) = 1$, either $p^\alpha \mid m$ and $(p^\alpha, n) = 1$ or $p^\alpha \mid n$ and $(p^\alpha, m) = 1$.   Since the theorem and the alternatives above are symmetrical in m and n, we shall treat only the first.   That is, suppose $p^\alpha \mid m$ and, necessarily, $(p^\alpha, n) = 1$. Then

$$p^\alpha \mid (xm + yn, mn) \Leftrightarrow p^\alpha \mid xm + yn$$

$$\Leftrightarrow p^\alpha \mid y$$

$$\Leftrightarrow p^\alpha \mid (y, m)$$

$$\Leftrightarrow p^\alpha \mid (x, n)(y, m).$$

10.4 Theorem.   The Euler φ-function is multiplicative.

Proof.   Let X be a set of coset representatives of the cosets in $\Phi(n)$.   We may as well take for X the positive integers less than n and relatively prime to n. Similarly define Y using m instead of n.   Then X has $\varphi(n)$ elements; Y has $\varphi(m)$.   We shall show that the set

$$W = \{xm + ny \mid x \in X, y \in Y\}$$

has $\varphi(m)\varphi(n)$ elements and is a complete set of representatives of the cosets in $\Phi(mn)$.   The theorem will then be proved.

Suppose $x \in X$ and $y \in Y$.   Then $(x, n) = (y, m) = 1$ so Lemma 10.3 implies that

$$(xm + yn, mn) = 1.$$

That is, W represents only cosets in $\Phi(mn)$.

Next, we show no two elements of W are congruent modulo mn.   It will follow that the elements of W represent different cosets in $\Phi(mn)$ and hence, *a fortiori*, no two elements of W are equal and W has $\varphi(m)\varphi(n)$ elements. Suppose

$$xm + yn \equiv x'm + y'n \pmod{mn} \tag{14}$$

for $x, x' \in X$ and $y, y' \in Y$. We must prove $x = x'$ and $y = y'$. Congruence (14) implies that

$$mn \mid (x - x')m + (y - y')n.$$

Apply Lemma 10.3:

$$mn \mid (x - x', n)(y - y', m).$$

Since m and n are relatively prime, it follows that

$$n \mid x - x' \qquad \text{and} \qquad m \mid y - y',$$

but distinct elements of X are incongruent modulo n, so x must equal x'. Similarly, $y = y'$.

Finally, we must show that W represents every coset in $\Phi(mn)$. Since $(m, n) = 1$, we can always solve the Diophantine equation

$$w = xm + ny$$

(Theorem 3.1). If w lies in an element of $\Phi(mn)$ then $(w, mn) = 1$. Lemma 10.3 then implies that

$$(x, n) = (y, m) = 1.$$

If we replace x by the element of X to which it is congruent modulo n and similarly replace y, we do not change $\bar{w}$ in $\mathbf{Z}_{mn}$. Therefore W represents every coset in $\Phi(mn)$.

10.5 Theorem $\varphi(n) = n \prod\limits_{\text{primes } p \mid n} \left(1 - \dfrac{1}{p}\right).$

Proof. First a remark on the notation. The symbol $\prod$ is to products as $\sum$ is to sums. That is, Theorem 10.5 may be restated as: if $p_1, \ldots, p_r$ are different primes and

$$n = p_1^{\alpha_1} \cdots p_r^{\alpha_r} \tag{15}$$

then

$$\varphi(n) = n\left(1 - \frac{1}{p_1}\right)\left(1 - \frac{1}{p_2}\right) \cdots \left(1 - \frac{1}{p_r}\right)$$

$$= n \prod_{i=1}^{r} \left(1 - \frac{1}{p_i}\right).$$

For example,

$$\varphi(12) = \varphi(2^2 \cdot 3) = 12(1 - \tfrac{1}{2})(1 - \tfrac{1}{3}) = 4$$

which checks with what we already know.

Now to prove the theorem. Suppose $n = p^\alpha$, a prime power. Then

$$\varphi(p^\alpha) = p^\alpha - p^{\alpha-1} = p^\alpha\left(1 - \frac{1}{p}\right)$$

(Theorem 10.1) so the theorem is true. Now suppose n any integer. Write n as in Eq. (15). Since φ is multiplicative,

$$\varphi(n) = \varphi(p_1^{\alpha_1}) \cdots \varphi(p_r^{\alpha_r})$$

$$= p_1^{\alpha_1}\left(1 - \frac{1}{p_1}\right) \cdots p_r^{\alpha_r}\left(1 - \frac{1}{p_r}\right)$$

$$= p_1^{\alpha_1} \cdots p_r^{\alpha_r}\left(1 - \frac{1}{p_1}\right) \cdots \left(1 - \frac{1}{p_r}\right)$$

$$= n \prod_{\text{primes } p_i | n} \left(1 - \frac{1}{p_i}\right).$$

11. PROBLEMS

11.1 Prove: If a is odd, then $a^2 \equiv 1(8)$. If a is even, then $a^2 \equiv 0(4)$. These facts help in Problems 6.1 and 6.4.

11.2 When does $ax \equiv bx(n)$ imply $a \equiv b(n)$?

11.3 Prove: If $a \equiv b(n_i)$ for mutually relatively prime integers $n_1, \ldots, n_r$, then $a \equiv b(n_1 \cdots n_r)$.

11.4 Prove the following theorem due to Lucas: If $\Phi(n)$ contains an element of order $n - 1$, then n is prime.

11.5 If $m \mid n$ then $n\mathbf{Z} \subset m\mathbf{Z}$, so there is a natural map $\pi_m{}^n: \mathbf{Z}/n\mathbf{Z} \to \mathbf{Z}/m\mathbf{Z}$. Use this map to describe the uniqueness assertion of Theorem 7.1.

11.6 Solve each of the following sets of simultaneous congruences:

$$\{x \equiv 1\ (2), \quad x \equiv 2\ (3), \quad x \equiv 3\ (5), \quad x \equiv 4\ (7)\}$$
$$\{y \equiv 1\ (9), \quad 2\ (5), \quad 1\ (7)\}$$
$$\{z \equiv 1\ (9), \quad 1\ (5), \quad 3\ (7)\}.$$

11.7 Prove this more general version of the Chinese remainder theorem: Let $n_1, \ldots, n_r$, $a_1, \ldots, a_r$, and $b_1, \ldots, b_r$ be integers such that $(n_i, n_j) = 1$ when

$i \neq j$ and such that each of the congruences $a_i x \equiv b_i(n_i)$ has a solution. Then these congruences have a simultaneous solution.

11.8 Prove this more general version of the Chinese remainder theorem: The congruences $x \equiv a_i(n_i)$, $i = 1, \ldots, r$, have a simultaneous solution if and only if $a_i \equiv a_j|(n_i, n_j)|$ whenever $i \neq j$.

11.9 Use Problem 11.5 to discuss the uniqueness of the solution produced by the Chinese remainder theorem.

11.10 An ideal-theoretic version of the Chinese remainder theorem. Let $I_1, \ldots, I_r$ be ideals in $\mathbf{Z}$ no two of which are contained in a common maximal ideal. Then for any integers $b_1, \ldots, b_r$ there is an $x \in \mathbf{Z}$ such that $x - b_j \in I_j$ for all j.

11.11 Prove that every finite integral domain is a field.

11.12 Divisibility tests. Write $m = a_k \vee a_{k-1} \vee \cdots \vee a_0$ for the integer m in base 10 notation. That is, $0 \leq a_j \leq 9$ and $m = a_k 10^k + \cdots + a_1 10 + a_0$. Prove:
(a) $m \equiv a_0$ (2),
(b) $m \equiv a_0 + a_1 + \cdots + a_k$ (3),
(c) $m \equiv a_1 \vee a_0 \equiv 2a_1 + a_0$ (4),
(d) $m \equiv a_0$ (5),
(e) $m \equiv a_2 \vee a_1 \vee a_0 \equiv 4a_2 + 2a_2 + a_0$ (8),
(f) $m \equiv a_0 + a_1 + \cdots + a_k$ (9),
(g) $m \equiv a_0$ (10),
(h) $m \equiv a_0 - a_1 + a_2 - \cdots + (-1)^k a_k$ (11).

11.13 Devise a test for divisibility by 7.

11.14 Use part (f) of Problem 11.12 to justify the well known method of "casting out nines" to check arithmetical computations.

11.15 Generalize the results of Problem 11.12 to find tests for the divisibility of m by q in terms of the digits of the expansion of m in the base b.

11.16 When is $\varphi(n)$ odd?

11.17 For which n do each of the following hold?

$$\varphi(2n) > \varphi(n), \qquad \varphi(2n) = \varphi(n), \qquad \varphi(2n) < \varphi(n).$$

11.18 Show that (x, y) is a multiplicative function of x for each fixed y.

11.19 Find all solutions to $\varphi(n) = 24$.

11.20 Prove

$$\sum_{\substack{k < n \\ (k,n) = 1}} k = \tfrac{1}{2}n\varphi(n).$$

11.21 Prove or disprove the following generalization of Lemma 10.3:

$$(m, n)(xm + yn, mn) = (x, n)(y, m).$$

11.22 Show that the ring $\mathbf{Z}_n$ is naturally isomorphic to the endomorphism ring of the additive group $(\mathbf{Z}_n, +)$. Deduce that $\Phi(n)$ is naturally isomorphic to the group of automorphisms of $(\mathbf{Z}_n, +)$.

11.23 For which integers n does $\mathbf{Z}_n$ contain a nontrivial idempotent, that is, an element $X \neq 0, 1$ for which $X^2 = X$?

11.24* Let $\mathfrak{A}_n(d)$ be the annihilator of d in $\mathbf{Z}_n$. That is,

$$\mathfrak{A}_n(d) = \{X \in \mathbf{Z}_n \mid X\bar{d} = 0\}$$
$$= \overline{\{x \in \mathbf{Z} \mid xd \equiv 0(n)\}}.$$

(a) Show $\mathfrak{A}_n(d) = \mathfrak{A}_n(d')$ if $d \equiv d'(n)$.
(b) Show $\mathfrak{A}_n(d) = \mathfrak{A}_n(xd)$ if $(x, n) = 1$.
(c) Prove that $\mathfrak{A}_n(d)$ is an ideal of the ring $\mathbf{Z}_n$.
(d) Part (c) shows $\mathfrak{A}_n(d)$ is itself a ring. Prove that it is a field if and only if (d, n) is a prime whose square does not divide n. (This is Advanced Problem No. 5469 in the *American Mathematical Monthly*, (1967) **74**, 208. The solution is in the same *Monthly*, **75**, (1968) 306.)

11.25 Let $\sigma(n)$ be the sum of the divisors of n. Thus $\sigma(1) = 1$, $\sigma(2) = 1 + 2 = 3$, $\sigma(6) = 1 + 2 + 3 + 6 = 12$, and so on. Prove
(a) σ is multiplicative.
(b) If p is prime then $\sigma(p^\alpha) = (p^{\alpha+1} - 1)/(p - 1)$.
(c) Derive a formula for $\sigma(n)$ analogous to the one for $\varphi(n)$ established in Theorem 10.5.

11.26 An integer is *perfect* if it is the sum of its proper divisors. For example, $6 = 1 + 2 + 3$ and $28 = 1 + 2 + 4 + 7 + 14$ are perfect; N is perfect if and only if $\sigma(N) = 2N$. Prove that $2^{n-1}M_n$ is perfect if M_n is a Mersenne prime (Problem 6.18). Euclid knew this fact.

11.27* Prove that if N is an even perfect number, then $N = 2^{n-1}M_n$ for some Mersenne prime M_n. Euler first proved that theorem. It is not known whether there is an odd perfect number. For more facts about these numbers see Martin Gardner's "Mathematical Games: A Short Treatise on the Useless Elegance of Perfect Numbers and Amicable Pairs," *Scientific American*, March, 1968, p. 121.

11.28 Prove: $\varphi(n)$ is a power of 2 if and only if

$$n = 2^\tau p_1 \cdots p_k$$

where the p_i are distinct Fermat primes (Problem 6.17). Gauss proved that a regular n-gon can be constructed with ruler and compass for just these values of n.

11.29 Let $\rho(n)$ be the number of integers $1 \leq k \leq n$ such that

$$(k, n) = (k + 1, n) = 1.$$

Show ρ is multiplicative, evaluate ρ at prime powers, and deduce

$$\rho(n) = n \prod_{p \mid n} \left(1 - \frac{2}{p}\right).$$

11.30* Arithmetic functions. Let $Z^+ = \{1, 2, 3, \ldots\}$ and R the set of functions from Z^+ to Z. Such a function is called arithmetic. Some useful members of R are φ, σ (defined in Problem 11.25), ρ (Problem 11.29), (x, y) for fixed y (Problem 11.18), and 0, ε, i, θ, τ defined by

$$0(n) = 0, \text{ all } n \in Z^+,$$

$$\varepsilon(n) = \begin{cases} 1 & \text{if } n = 1 \\ 0 & \text{otherwise,} \end{cases}$$

$$i(n) = 1, \text{ all } n \in Z^+,$$

$$\theta(n) = n, \text{ all } n \in Z^+,$$

$$\tau(n) = \text{the number of positive divisors of } n.$$

Define $+$ and $*$ in R by

$$(\alpha + \beta)(n) = \alpha(n) + \beta(n)$$

$$(\alpha * \beta)(n) = \sum_{d \mid n} \alpha(d)\beta\left(\frac{n}{d}\right).$$

(a) Show $\tau = i * i$, $\sigma = i * \theta$, $\theta = \varphi * i$, $\varphi * \tau = \sigma$ (see (d)).
(b) Prove that $(R, +, *)$ is an integral domain in which 0 is the additive and ε the multiplicative identity.
(c) The function α is a unit in R if and only if $\alpha(1) = \pm 1$.
(d) The set of nonzero multiplicative arithmetic functions is a subgroup of the group of units of R.
(e) The function $\mu = i^{-1}$ is called the *Möbius function*. Deduce the *Möbius inversion formula*

$$\alpha(n) = \sum_{d \mid n} \beta(d) \Rightarrow \beta(n) = \sum_{d \mid n} \alpha(d)\mu\left(\frac{n}{d}\right) \tag{16}$$

from the implication

$$\alpha = \beta * i \Rightarrow \beta = \alpha * \mu.$$

(f) Show that

$$\mu(n) = \begin{cases} 1 & \text{if } n = 1 \\ 0 & \text{if } m^2 \mid n \text{ for some } m > 0 \\ (-1)^k & \text{if } n \text{ is the product of } k \text{ distinct primes.} \end{cases}$$

(g) Show $\theta^{-1}(n) = \theta(n)\mu(n)$, all n, and $\sum_{d \mid n} \mu(d) = 0$ $(n > 1)$.

(h) Take (f|) as the definition of μ, and prove the inversion formula (16) directly, that is, without (b).

3

Polynomials

In this chapter we shall prove Wilson's theorem and a theorem of Fermat's which says that a prime congruent to 1 modulo 4 is a sum of two squares. These results and others we shall use later follow easily from facts about the algebra of polynomials.

12. THE ALGEBRA OF POLYNOMIALS

Let R be a commutative ring with identity $1 \neq 0$; write $R[x]$ for the ring of polynomials in the variable x with coefficients in R. The subring of $R[x]$ consisting of "constant" polynomials is naturally identified with R.

Suppose $r \in R$. Then the map $\sigma_r : R[x] \to R$ obtained by "substituting" r for x in each polynomial is a ring homomorphism. We follow the customary convenient practice and write $f(r)$ for $\sigma_r(f)$. If $s \in R$ is thought of as the constant polynomial $s \in R[x]$, then $s(r) = s$ for all r; r is a *root* of f when $f(r) = 0$.

If S is a ring and $\rho : R \to S$ is a ring homomorphism, then ρ extends naturally to the homomorphism from $R[x]$ to $S[x]$ obtained by applying ρ to the coefficients of each polynomial. We shall also call this extended homomorphism ρ. If $f \in R[x]$ and $r \in R$, then $\rho(f(r)) = \rho(f)(\rho(r))$.

In particular, the function assigning to each polynomial with integral coefficients the polynomial with the same coefficients reduced modulo n is a homomorphism from $\mathbf{Z}[x]$ to $\mathbf{Z}_n[x]$. When $f \in \mathbf{Z}[x]$ we write $\bar{f}$ for its image in $\mathbf{Z}_n[x]$; $\bar{f}$ is the polynomial whose coefficients are the coefficients of f modulo n. If f has an integral root r, then clearly $\bar{r}$ is a root of $\bar{f}$, but the converse is false. For example, 2 is a root of $x^2 + 1$ in $\mathbf{Z}_5$, but $x^2 + 1$ has no root in $\mathbf{Z}$. In fact, $\bar{f}$ has a root in $\mathbf{Z}_n$ if and only if the Diophantine equation

$$f(x) = ny$$

has a solution. When this happens we say f has a root modulo n.

Let $f \in \mathbf{Z}[x]$. Since $f(m) \equiv 0(f(m))$, f has a root modulo $f(m)$ for every $m \in \mathbf{Z}$. Thus every nonconstant polynomial has roots modulo n for infinitely many integers n. We are about to prove more.

12.1 Theorem. Every nonconstant polynomial $f \in \mathbf{Z}[x]$ has roots modulo p for infinitely many primes p.

Proof. Let

$$f(x) = a_n x^n + \cdots + a_1 x + a_0$$

where $n > 0$ and $a_n \neq 0$. If f has an integral root k, then $f(k) = 0$ so $f(k) \equiv 0(p)$ for every prime p, and the theorem is true for f. Therefore, assume f has no integral root. In particular, $f(0) = a_0 \neq 0$. Suppose that for primes $p_1, \ldots, p_r$, $f(x) \equiv 0(p_i)$ has a solution. We shall find another prime with that property. Consider the polynomial g in $\mathbf{Z}[x]$ defined by

$$f(p_i \cdots p_r a_0 x) = a_0 (c_n x^n + \cdots + c_1 x + 1)$$
$$= a_0 g(x).$$

Each of the coefficients $c_1, \ldots, c_n \neq 0$ of g is divisible by $p_1 \cdots p_r$. Since g is not constant, there is an integer x_0 such that $g(x_0) \neq \pm 1$. Let p be a prime factor of $g(x_0)$. Since $g(x_0) \equiv 1(p_i)$, p is not one of the primes $p_1, \ldots, p_r$. But $p \mid a_0 g(x_0) = f(p_1 \cdots p_r a_0 x_0)$, so f has a root modulo p.

We are now ready to settle a question left open in Section 5: the infinitude of the primes congruent to 1 modulo 4.

12.2 Lemma. If p is an odd prime and $x^2 + 1 \equiv 0(p)$ has a solution, then $p \equiv 1(4)$.

Proof. Let k be a root of $x^2 + 1$ modulo p.   Then $k^2 \equiv -1(p)$ so

$$1 \equiv k^{p-1} \qquad (p) \qquad \text{(Fermat's theorem)}$$
$$\equiv (k^2)^{(p-1)/2} \quad (p)$$
$$\equiv (-1)^{(p-1/2)} \, (p).$$

Thus $(p - 1)/2$ is even and hence $p \equiv 1(4)$.

12.3 Corollary. There are infinitely many primes congruent to 1 modulo 4.

Proof. Each of the infinitely many odd primes modulo whichever $x^2 + 1$ has a root is congruent to 1 modulo 4.

The converse to Lemma 12.2 is true.   We shall prove it in the next section using some properties of $\mathbf{Z}_p[x]$ which follow from the fact that $\mathbf{Z}_p$ is a field.

13. WILSON'S THEOREM

Let F be a field.   Then $F[x]$ possesses a Euclidean norm and hence is a unique factorization domain (Appendix 1, Example 9, and Theorem 16). $a \in F$ is a root of $f \in F[x]$ if and only if $(x - a) \,|\, f$.   Thus a polynomial of degree n has at most n roots in F (Appendix 1, Theorem 17, and Corollary 18).   Since $\mathbf{Z}_p$ is a field when p is prime (Theorem 9.5), these remarks apply to $\mathbf{Z}_p[x]$.

13.1 Wilson's Theorem. If p is prime, then $(p - 1)! \equiv -1(p)$.

Proof. Consider the polynomial $f(x) = x^{p-1} - 1$ in $\mathbf{Z}_p[x]$.   Fermat's theorem implies that the $p - 1$ elements $1, 2, \ldots, p - 1$ of $\mathbf{Z}_p$ are all roots of f, which is of degree $p - 1$.   Then the introductory remarks above show

$$x^{p-1} - 1 = (x - 1)(x - 2) \cdots (x - (p - 1))$$

in $\mathbf{Z}_p[x]$.   Now substitute $x = 0$ in $\mathbf{Z}_p$:

$$-1 \equiv (-1)^{p-1}(p - 1)! \equiv (p - 1)! \; (p)$$

since either $p - 1$ is even or $p = 2$, in which case $-1 \equiv 1(p)$.

13.2 Theorem. $x^2 + 1$ has a root modulo the odd prime p, or equivalently, $x^2 \equiv -1(p)$ has a solution, if and only if $p \equiv 1(4)$.

Proof. "Only if" is Lemma 13.2, so suppose $p \equiv 1(4)$. Then $(p-1)/2$ is even. Now

$$1 \cdot 2 \cdot 3 \cdots (p-2)(p-1) \equiv -1 \ (p) \qquad (1)$$

(Wilson's theorem) and

$$-r \equiv p - r \ (p)$$

for $r = 1, 2, \ldots, (p-1)/2$. Thus rearranging the factors in (1) yields

$$-1 \equiv 1 \cdot (-1) \cdot 2 \cdot (-2) \cdots \frac{p-1}{2} \left(-\frac{p-1}{2}\right) (p)$$

$$\equiv (-1)^{(p-1)/2} \left(\frac{p-1}{2}!\right)^2 (p)$$

$$\equiv \left(\frac{p-1}{2}!\right)^2 (p).$$

The two square roots of -1 in $\mathbf{Z}_p$, and hence the roots of $x^2 + 1$ modulo p, are thus $\pm \left(\dfrac{p-1}{2}\right)!$

14. THE DIOPHANTINE EQUATION $x^2 + y^2 = p$

Which primes p are the sums of two squares; that is, which primes are representable in the sense of Section 1? Fermat knew the answer we are about to give. Since $2 = 1^2 + 1^2$, the only even prime is disposed of. Henceforth let p be an odd prime.

14.1 Lemma. If $x^2 + y^2 = p$, then $(x, p) = (y, p) = 1$.

Proof. If $p \mid x$, then $p^2 \mid x^2$, so $p \mid y^2$. Since p is prime, this implies $p \mid y$, so $p^2 \mid y^2$ and hence $p^2 \mid p$, which is absurd. Thus $p \nmid x$, so $(x, p) = 1$. Similarly, $(y, p) = 1$.

14.2 Lemma. If $x^2 + y^2 = p$, then $p \equiv 1(4)$.

Proof. Reduction modulo p yields

$$x^2 \equiv -y^2 \ (p)$$

and thus

$$(xy^{-1})^2 \equiv -1 \ (p)$$

where y^{-1} is the inverse of y in the group $\Phi(p)$. Lemma 12.2 then implies $p \equiv 1(4)$.

Here is a second proof. Suppose $x^2 + y^2 = p$. Since p is odd, x and y have opposite parity. Then one of x^2 or y^2 is divisible by 4; the other is congruent to 1 modulo 4 (Problem 11.1. Also compare Problem 1.1).

The second proof is simpler than the first, but the first can be made to run backwards to prove the converse of Lemma 14.2. Suppose p is a prime congruent to 1 modulo 4. Then there is an integer z such that $z^2 \equiv -1(p)$ (Theorem 13.2). We shall try to write $z = xy^{-1}$ in $\Phi(p)$ for small values of $|x|$ and $|y|$, for then

$$z^2 \equiv (xy^{-1})^2 \equiv -1 \ (p)$$

implies

$$x^2 \equiv -y^2 \ (p)$$

so

$$x^2 + y^2 = kp$$

for some integer k. If $|x|$ and $|y|$ are small enough, k will have to be 1. This motivates the next theorem and its application to the problem at hand. Before stating it we shall make explicit the obvious combinatorial fact on which its proof depends, Dirichlet's famous "pigeonhole principle".

14.3 Lemma. If n objects are filed in fewer than n pigeonholes, then at least two are filed together. More mathematically, though less vividly: If an n element set is a union of $m < n$ of its subsets, then at least one of those subsets has more than one element.

14.4 Theorem (Thue). Suppose n is not a perfect square and $z \in \mathbf{Z}$ Then the congruence

$$xz \equiv y \ (n)$$

has a solution $\langle x, y \rangle$ for which $|x|$ and $|y|$ are both less than $\sqrt{n}$ but are not both zero.

Proof. Consider all the integers $xz - y$ such that x and y are nonnegative and less than $\sqrt{n}$. Let m be the least integer greater than $\sqrt{n}$. Then there are m choices for x and m for y. Since $m^2 > n$ and $\mathbf{Z}_n$ has n elements, there must be two unequal pairs $\langle x, y \rangle$ and $\langle x', y' \rangle$ such that $\overline{(xz - y)} = \overline{(x'z - y')}$ in $\mathbf{Z}_n$. (This is the pigeonhole principle.) Then

$$(x - x')z \equiv y - y' \ (n).$$

Since $|x - x'|$ and $|y - y'|$ are both less than $\sqrt{n}$ and are not both zero, we are done.

14.5 Theorem (Fermat). An odd prime is the sum of two squares if and only if it is congruent to 1 modulo 4.

Proof. "Only if" is Lemma 14.2, so suppose $p \equiv 1(4)$. Let z satisfy

$$z^2 \equiv -1 \ (p) \tag{2}$$

(Theorem 2.2).

Choose x and y satisfying the conclusions of Thue's theorem (14.4). Then (2) and $xz \equiv y(p)$ imply

$$-x^2 \equiv x^2 z^2 \equiv y^2 \ (p)$$

so

$$x^2 + y^2 = kp$$

for some integer k.

Since $|x| < \sqrt{p}$ and $|y| < \sqrt{p}$, $x^2 + y^2 < 2p$ so $k < 2$. Since k is a positive integer, $k = 1$ and

$$x^2 + y^2 = p \tag{3}$$

as desired.

This is our first encounter with a member of the class of Diophantine equations

$$x^2 - my^2 = n.$$

We shall have a lot to say about this equation for some small values of $|m|$ and arbitrary n.

We close this section by studying the Diophantine equation

$$x^2 - y^2 = p. \tag{4}$$

The seemingly innocuous change of sign makes the problem trivial. Suppose p is a positive prime and $x > y > 0$. Then since $x^2 - y^2 = (x - y)(x + y)$, Eq. (4) implies

$$x - y = 1 \quad \text{and} \quad x + y = p$$

so

$$x = \frac{p + 1}{2} \quad \text{and} \quad y = \frac{p - 1}{2}.$$

Thus the prime p is a difference of integral squares if and only if it is odd. In fact, we have discovered a little more.

14.6   Theorem. Every odd integer is the difference of two consecutive squares.

Proof. If n is odd, then $(n \pm 1)/2 \in \mathbf{Z}$ and

$$n = \left(\frac{n + 1}{2}\right)^2 - \left(\frac{n - 1}{2}\right)^2.$$

Problem 15.10 considers the representation of an even integer as a difference of squares.

15.   PROBLEMS

15.1   Let F be a field. Show that the units in $F[x]$ are the constant polynomials.

15.2   If $h \in \mathbf{Z}_p[x]$, we may regard h as a function from $\mathbf{Z}_p \to \mathbf{Z}_p$ by substituting for x the elements of $\mathbf{Z}_p$. Show that for f and $g \in \mathbf{Z}[x]$ the following are equivalent:

(a) $f(n) \equiv g(n)(p)$ for all $n \in \mathbf{Z}$.
(b) $\bar{f} \equiv \bar{g}(x^p - x)$ in $\mathbf{Z}_p[x]$. That is, $x^p - x \mid \bar{f} - \bar{g}$ in $\mathbf{Z}_p[x]$.
(c) $\bar{f}$ and $\bar{g}$ yield the same function from $\mathbf{Z}_p \to \mathbf{Z}_p$.

Show by example that none of these conditions implies $\bar{f} = \bar{g}$.

15.3*   Prove that there are infinitely many prime (that is, irreducible) polynomials in $\mathbf{Z}_p[x]$. Write down the prime polynomials of degree less than or equal to 3 in $\mathbf{Z}_2[x]$ and $\mathbf{Z}_3[x]$.

15.4   Show that n is prime if and only if every linear polynomial in $\mathbf{Z}_n[x]$ has at most one root in $\mathbf{Z}_n$.

15.5   Let $f \in \mathbf{Z}[x]$ be monic, that is, suppose its leading coefficient is 1. Show that the only rational roots of f are integers.

15.6 Deduce from Problem 15.5 that $m^{1/n}$ is irrational unless m is the nth power of an integer.

15.7* Prove Wilson's theorem by counting the number of p-Sylow subgroups of the symmetric group on p symbols.

15.8 Prove the converse of Wilson's theorem.

15.9* The kth elementary symmetric function of n variables, $S_k{}^n$, is defined by

$$S_k{}^n(x_1, \ldots, x_n) = \sum_{1 \le i_1 < \cdots < i_k \le n} x_{i_1} \cdots x_{i_k}.$$

Then

$$\prod_{j=1}^{n} (X - x_j) = \sum_{k=0}^{n} (-1)^{n-k} S_k{}^n(x_1, \ldots, x_n) X^k$$

is true in $F[X]$ for every field F. The *fundamental theorem on symmetric functions* says that any polynomial in n variables with coefficients in F which is invariant under all permutations of its arguments is a polynomial in the elementary symmetric functions. For example,

$$x^2 + y^2 + z^2 = (S_1{}^3(x, y, z))^2 - 2S_2{}^3(x, y, z).$$

Investigate $S_k{}^n(1, 2, \ldots, p-1)$ modulo p when p is an odd prime. Investigate $1^2 + 2^2 + \cdots + (p-1)^2$ in $\mathbf{Z}_p$.

15.10 Show that an even integer is a difference of squares if and only if it is doubly even, that is, divisible by 4.

15.11 The argument preceding Theorem 14.6 shows that the representation of a prime as a difference of squares is unique. However

$$15 = 8^2 - 7^2 = 4^2 - 1^2.$$

We can count the number of solutions to the Diophantine equation

$$x^2 - y^2 = n. \tag{5}$$

Let

$$n = 2^\alpha p_1^{\alpha_1} \cdots p_k^{\alpha_k}$$

be the factorization of n as a product of powers of distinct primes. Let

$$N = (\alpha - 1)(\alpha_1 + 1) \cdots (\alpha_k + 1).$$

Prove that Eq. (5) has $N/2$ positive solutions if N is even and $(N+1)/2$ if N is odd.

15.12* Which integers can be written as a sum of consecutive odd positive integers? Of two or more consecutive odd positive integers?

15.13* Answer the questions posed in Problem 15.12 when the word "odd" is deleted.

15.14* Reread Section 1 and Problems 6.1, 6.2, and 6.3. Theorem 14.5 and Problem 15.11 may suggest new conjectures on representable integers and the number of ways to represent them.

4

The Group of Units of $\mathbb{Z}_n$

We shall show in this chapter that Φ, regarded as a group-valued function of n, is multiplicative. That fact together with an analysis of the structure of $\Phi(n)$ when n is a power of a prime will allow us to answer classical questions about the congruence $x^\alpha \equiv m(n)$.

16. DECIMAL EXPANSIONS

In this section we shall investigate the form of the decimal expansion of $1/n$; the questions raised by that investigation motivate the subsequent discussion of the group $\Phi(n)$.

First we shall do some arithmetic to provide ourselves with numerical examples.

$$\frac{1}{7} = 0.\overline{142857}. \tag{1}$$

The digits under the bar are to be repeated, that is,

$$\frac{1}{7} = 0.142857 \quad 142857 \quad 142857 \ldots.$$

We shall ignore all questions about the convergence of the infinite decimals we use. Any question the reader wishes to raise he must answer for himself.
Equation (1) follows from

$$
\begin{array}{r}
.142857 \\
7\overline{)1.000000} \\
-7 \\
\hline
30 \\
-28 \\
\hline
20 \\
-14 \\
\hline
60 \\
-56 \\
\hline
40 \\
-35 \\
\hline
50 \\
-49 \\
\hline
\mathbf{1}
\end{array}
$$

The sequence of "remainders," which appear in bold face, is

$$\overline{1, 3, 2, 6, 4, 5} = 1, 3, 2, 6, 4, 5, \qquad 1, 3, 2, 6, 4, 5, \ldots.$$

Similar computations show that

$$\frac{1}{80} = 0.0125\bar{0}, \tag{2}$$

where the remainders are $1, 10, 20, 40, \bar{0}$; that

$$\frac{1}{13} = 0.\overline{076923}, \tag{3}$$

where the remainders are $\overline{1, 10, 9, 12, 3, 4}$; and that

$$\frac{1}{88} = 0.011\overline{36}, \tag{4}$$

where the remainders are $1, 10, 12, \overline{32, 56}$.

Now fix a positive integer n. Suppose that

$$\frac{1}{n} = 0.a_1 a_2 \ldots, \tag{5}$$

where $0 \le a_i \le 9$, and that

$$1, r_1, r_2, \ldots, \tag{6}$$

where $0 \le r_i < n$, is the sequence of remainders which occurs in the long division algorithm. We wish to consider the remainders r_i both as integers and as elements of $\mathbf{Z}_n$. The rule "bring down the next 0" shows that

$$10r_i = a_i n + r_{i+1}. \tag{7}$$

Therefore

$$r_{i+1} \equiv 10r_i \ (n). \tag{8}$$

Since $r_0 = 1$, (8) implies

$$r_{i+1} \equiv 10^i \ (n). \tag{9}$$

The infinite sequence (6) of remainders lies in the finite set $\{0, 1, \ldots, n-1\}$, so there must be a first repetition

$$r_\mu = r_{\mu+\lambda}. \tag{10}$$

Then the long division algorithm implies that the sequence of remainders is just

$$1, r_1, \ldots, \overline{r_\mu, r_{\mu+1}, \ldots, r_{\mu+\lambda-1}} \tag{11}$$

and the corresponding decimal fraction is

$$\frac{1}{n} = 0.a_1 \ldots a_\mu \overline{a_{\mu+1} \ldots a_{\mu+\lambda}}. \tag{12}$$

We call λ the *period* of the expansion in Eq. (12); that expansion is *purely periodic* if and only if $\mu = 0$, or, equivalently, the first repetition in the sequence (11) is $r_\lambda = 1$. The expansion terminates if and only if $r_\mu = 0$. In that case all succeeding remainders will be 0. We wish to discover how μ and λ depend on n. To do so we review some elementary group theory.

Let G be a finite group with identity e, and g an element of G. Then the first repetition in the sequence

$$e, g, g^2, \ldots \tag{13}$$

is of the form $g^\lambda = e$ and λ is the order of g. The map $\Psi : \mathbf{Z} \to G$ given by $\Psi(n) = g^n$ is then a homomorphism with kernel $\lambda \mathbf{Z}$, so that it may be regarded as an isomorphism between the additive group $\mathbf{Z}_\lambda = \mathbf{Z}/\lambda\mathbf{Z}$ and the subgroup $\{e, g, \ldots, g^{\lambda-1}\}$ of G spanned by g.

16.1 Theorem. Each of the following three groups of statements consists of equivalent statements. For any particular positive integer n exactly one of these groups consists of true assertions.

I (a) The decimal expansion for $1/n$ is purely periodic.
  (b) $(10, n) = 1$; that is, neither 2 nor 5 divides n.
  (c) For some $\lambda > 0$, $r_\lambda = 1$.
II (a) The decimal expansion for $1/n$ terminates.
  (b) For some μ, $n \mid 10^\mu$; that is, n has no prime factors other than 2 or 5.
  (c) For some μ, $r_\mu = 0$.
III (a) The decimal expansion for $1/n$ is not purely periodic and does not terminate.
  (b) 2 or 5 and some third prime divide n.
  (c) For $i > 0$, r_i is never 0 or 1.

Proof. Let n be a positive integer. Elementary logic shows just one of I(a), II(a), or III(a) and just one of I(c), II(c), and III(c) is true. A little reflection shows just one of I(b), II(b), or III(b) is true. Thus to prove the theorem it suffices to show I(a) $\Leftrightarrow$ I(b) $\Leftrightarrow$ I(c) and II(a) $\Leftrightarrow$ II(b) $\Leftrightarrow$ II(c).

We treat case I first. Suppose $(10, n) = 1$ (I(b)). Then $\overline{10} \in \Phi(n)$ (Theorem 9.2). Congruence (9) now implies that sequence (6) of remainders is just the sequence (13) when $g = \overline{10}$ in the group $\Phi(n)$. Therefore the first repetition is $r_\lambda = 1$ and I(a) is true.

Suppose I(a) true. Then $r_\lambda = r_0 = 1$, so I(c) follows. Finally, suppose I(c) true. Then

$$1 \equiv 10^\lambda \equiv 10 \cdot 10^{\lambda-1}(n).$$

That is, $\overline{10}$ is invertible in $\mathbf{Z}_n$, so $(10, n) = 1$ (I(b)) (Theorem 9.2).

Case II is simpler. We show II(a) $\Rightarrow$ II(b) $\Rightarrow$ II(c) $\Rightarrow$ II(a). If the decimal expansion for n is

$$\frac{1}{n} = 0. \, a_1 \ldots a_\mu \bar{0}$$

then $10^\mu/n$ is an integer, so $n \mid 10^\mu$. If $n \mid 10^\mu$, then (9) shows $r_\mu = 0$. If $r_\mu = 0$, then for all $k \geq 0$, $r_{\mu+k}$ and hence $a_{\mu+k+1}$ is zero.

In our examples above, 7 and 13 are covered by Case I, 80 by Case II, and 88 by Case III.

For the remainder of this section we shall restrict our attention to Case I.

16.2   Corollary.   Suppose $(10, n) = 1$.   Let Λ be the subgroup of $\Phi(n)$ generated by $\overline{10}$.   Then the period $\lambda(n)$ of the decimal expansion of $1/n$ is the order of Λ and hence divides $\varphi(n)$.

The only special significance of 10 in this section is the fact that we have 10 fingers and so write numbers decimally.   The methods we used really prove more than we have so far made explicit.   The following theorem states a consequence of Case I for expansions to any base.

16.3   Theorem.   Suppose $(m, n) = 1$.   Let Λ_m be the subgroup of $\Phi(n)$ generated by $\overline{m}$.   Then the period $\lambda_m(n)$ of the expansion of $1/n$ in the base m is the order of Λ_m and hence divides $\varphi(n)$.

We shall continue to write $\lambda(n)$ for $\lambda_{10}(n)$.   A question commonly asked is: For which n does $\lambda(n) = \varphi(n)$?   The integer 7 enjoys this property; 13 does not.   The question is equivalent to: For which n is $\Phi(n)$ cyclic with $\overline{10}$ as a generator?

The bulk of this chapter is devoted to the structure of $\Phi(n)$; when we are done we shall know when $\Phi(n)$ is cyclic.   The problem of deciding whether 10 happens to be a generator is unsolved.   For example, we shall see that the 12 element group $\Phi(13)$ is cyclic, though we know that the order of 10 in that group is only 6.   In general $\Phi(p)$ is always cyclic when p is prime. Part of a conjecture due to Artin asserts that 10 generates $\Phi(p)$ for infinitely many primes p.

We close this section with some remarks on the decimal expansion of k/n. Suppose $(10, n) = 1$ and that k/n is in lowest terms, so that $(k, n) = 1$.   If k happens to be in Λ, then it is just one of the remainders which appeared when we worked out the decimal expansion of $1/n$.   Then the decimal expansion of k/n is purely periodic; its block of digits is a cyclic permutation of the block for n.   For example,

$$10/13 = 0.\overline{769230}$$

and

$$9/13 = 0.\overline{692307}.$$

If $k \notin \Lambda$ we must begin again. Then the successive remainders in the division algorithm for k/n are just the numbers $\overline{10^i k}$. Thus they exhaust a coset of Λ in $\Phi(n)$. Therefore, k/n too has a purely periodic expansion with period $\lambda(n)$, and the cyclic permutations of its block of digits determine the expansions of the other elements of the Λ coset of k in $\Phi(n)$.

17. CYCLIC GROUPS

In this section we shall find criteria for deciding when a group is cyclic or a product of cyclic groups. Using them we shall be able to prove that $\Phi(p^\alpha)$ is cyclic when p is an odd prime and that $\Phi(n)$ is a product of cyclic groups in a useful way when n is divisible by several primes.

17.1 Definition. Let g generate a cyclic group G of order n. For $a \in G$ let the *index of* a *relative to* g be the least nonnegative integer m for which $a = g^m$.

Write $m = \mathrm{ind}_g(a)$. Then $0 \le \mathrm{ind}_g(a) \le n - 1$, and

$$g^{\mathrm{ind}\, g^{(a)}} = a. \tag{14}$$

When we regard the index as a map

$$\mathrm{ind}_g : G \to \mathbf{Z}_n, +$$

from G to the additive group of the ring $\mathbf{Z}_n$, then it is a group isomorphism. That is, to multiply two elements of G simply add their indices modulo n. The index of e is 0, the index of g is 1.

The index map should be thought of as a logarithm to the base g, for it turns multiplication in G into addition in $\mathbf{Z}_n$:

$$\mathrm{ind}_g(a + b) \equiv \mathrm{ind}_g(a) + \mathrm{ind}_g(b) \,(n).$$

Choosing a generator g as a base for the indices is equivalent to choosing a particular isomorphism of G with $\mathbf{Z}_n$, +.

These introductory remarks show that the study of finite cyclic groups is equivalent to the study of the groups $\mathbf{Z}_n$, +. For the remainder of this section $\mathbf{Z}_n$ will mean simply the additive group of the ring $\mathbf{Z}_n$.

17.2 Lemma. The order of $a \in \mathbf{Z}_n$ is $n/(a, n)$.

Proof. Remember that we are discussing $\mathbf{Z}_n$ as an additive group. Thus the order of a is the least positive k for which $ka \equiv 0(n)$. But Theorem 7.1

tells us how to find all such k; 0 is a solution and the solutions are unique modulo $n/(a, n)$. Therefore $n/(a, n)$ is the least positive solution. Note that this result is valid even for $a = 0$ since the order of 0 is 1 and $(0, n) = n$.

17.3   Corollary.   The element a generates Z_n if and only if $a \in \Phi(n)$, so that a cyclic group of order n has $\varphi(n)$ generators.

Proof.   The element a generates $Z_n \Leftrightarrow$ the order of a is $n \Leftrightarrow (a, n) = 1$ (Lemma 17.2) $\Leftrightarrow a \in \Phi(n)$.

17.4   Lemma.   If a group G contains an element of order d, then it contains at least $\varphi(d)$ of them.

Proof.   Suppose $g \in G$ has order d.   Then the cyclic subgroup of G generated by g has d elements and hence $\varphi(d)$ generators (Corollary 17.3) all of which are elements of G of order d.

17.5   Corollary.   The additive group Z_n has exactly $\varphi(d)$ elements of order d for each divisor d of n.

Proof.   Let $d \mid n$ and let $\Psi(d)$ be the number of elements of order d in Z_n. Since the order of n/d is d, $\Psi(d) \geq 1$.   Then $\Psi(d) \geq \varphi(d)$ (Lemma 17.4). But every element of Z_n has order d for some $d \mid n$, so

$$n = \sum_{d \mid n} \Psi(d) \geq \sum_{d \mid n} \varphi(d) = n. \qquad (Eq.\ (10),\ \text{Chapter } 2)$$

This can happen only if $\Psi(d) = \varphi(d)$ for all d.

A similar argument proves a kind of converse to this corollary.

17.6   Theorem.   Let G be a group of order n.   If G has at most $\varphi(d)$ elements of order d for each divisor d of n, then G is cyclic.

Proof.   Let d and $\Psi(d)$ be as in the proof of Corollary 17.5.   Then by hypothesis $\Psi(d) \leq \varphi(d)$ for all d, so

$$n = \sum_{d \mid n} \Psi(d) \leq \sum_{d \mid n} \varphi(d) = n.$$

This implies $\Psi(d) = \varphi(d)$ for all d.   In particular, $\Psi(n) = \varphi(n) \neq 0$, so there is an element of order n.   Thus G is cyclic.

We shall be particularly interested later in products of cyclic groups. The topic is quite general because every finite abelian group is such a product, but we make no use of that fact.

Suppose G and H are groups. Then the set $G \times H$ of ordered pairs $\langle g, h \rangle$ with $g \in G$ and $h \in H$ is a group when equipped with the multiplication

$$\langle g, h \rangle \langle g', h' \rangle = \langle gg', hh' \rangle.$$

The group $G \times H$ is called the *product* of the groups G and H. The definition of $G = G_1 \times \cdots \times G_n$ for any finite sequence of groups is straightforward; G will be abelian if and only if every factor G_i is abelian.

17.7 Lemma. The order of $\langle g, h \rangle$ in $G \times H$ is the least common multiple of the orders of g in G and h in H.

Proof. The identity element e of $G \times H$ is $\langle e_G, e_H \rangle$, so $\langle g, h \rangle^n = e$ if and only if $g^n = e_G$ in G and $h^n = e_H$ in H. That happens if and only if both the order of g and the order of h divide n. The lemma follows.

17.8 Corollary. The group $Z_m \times Z_n$ is cyclic if and only if $(m, n) = 1$.

Proof. Lemmas 17.2 and 17.7 imply that the order d of $\langle a, b \rangle$ in $Z_m \times Z_n$ is

$$d = \text{l.c.m.} \left\{ \frac{m}{(a, m)}, \ \frac{n}{(b, n)} \right\}$$

which divides l.c.m.$\{m, n\} = mn/(m, n)$. If $(m, n) > 1$, then no d can be as large as mn, which is the order of $Z_m \times Z_n$, so that group is not cyclic. Conversely, if $(m, n) = 1$, then $\langle 1, 1 \rangle$ has order mn and so generates $Z_m \times Z_n$.

17.9 Theorem. If $G \times H$ is cyclic, then G and H are cyclic.

Proof. Let $\langle g, h \rangle$ generate $G \times H$. Suppose G has order m and H order n. Then

$$mn = \text{order of } \langle g, h \rangle \text{ in } G \times H$$

$$= \text{l.c.m.} \{ \text{order of } g, \text{ order of } h \}$$

which is a divisor of mn. Thus the order of g is m, and the order of h is n.

17.10 Theorem. Let G be an abelian group. Suppose $g_1, \ldots, g$, are elements of G of orders $n_1, \ldots, n_r$ respectively. Then the map

$$\tau : \mathbf{Z}_{n_1} \times \cdots \times \mathbf{Z}_{n_r} \to G$$

given by

$$\tau(\langle k_1, \ldots, k_r \rangle) = g_1^{k_1} \cdots g_r^{k_r} \tag{15}$$

is a group homomorphism onto the subgroup G' of G generated by $g_1, \ldots, g_r$.

Proof. If the order of g is n, $g^k g^l = g^{k+l}$ where the addition in the exponent is performed modulo n. Thus

$$\begin{aligned}
\tau(\langle k_1, \ldots, k_r \rangle \langle l_1, \ldots, l_r \rangle) &= g_1^{k_1} \cdots g_r^{k_r} g_1^{l_1} \cdots g_r^{l_r} \\
&= g_1^{k_1 + l_1} \cdots g_r^{k_r + l_r} \\
&= \tau(\langle k_1 + l_1, \ldots, k_r + l_r \rangle)
\end{aligned}$$

where addition in the ith place is modulo n_i. Therefore τ is a group homomorphism. Since

$$\tau(\langle 0, \ldots, 1, \ldots, 0 \rangle) = g_i$$

when the 1 is in the ith place, the image of τ contains each g_i and thus is G'.

The proof above depends in a subtle way on the notation we used. We implicitly invoked the identification of $\mathbf{Z}_{n_i}$ with $\{1, 2, \ldots, n_i\}$ in order to define τ by Eq. (15) and then conveniently ignored the identification for the rest of the proof. The argument is however essentially correct. Rather than make it more pedantic by resolving the ambiguities of the notation we shall give another, more abstract version.

Write $\mathbf{Z}^r$ for $\mathbf{Z} \times \cdots \times \mathbf{Z}$ (r times). The map

$$T: \mathbf{Z}^r \to G$$

given by

$$T(\langle k_1, \ldots, k_r \rangle) = g_1^{k_1} \cdots g_r^{k_r}$$

is clearly a group homomorphism, and no notational ambiguity besets its definition. The kernel of T contains the subgroup $H = n_1 \mathbf{Z} \times \cdots \times n_r \mathbf{Z}$ of $\mathbf{Z}^r$, so the fundamental theorem of group homomorphisms implies that there is a homomorphism

$$\tau : \mathbf{Z}^r / H \to G \qquad (16)$$

such that $T = \tau \circ \pi$, where π is the natural projection of $\mathbf{Z}^r$ onto $\mathbf{Z}^r / H$. When $\mathbf{Z}^r / H$ and $\mathbf{Z}_{n_1} \times \cdots \times \mathbf{Z}_{n_r}$ are identified in the obvious way, the homomorphism τ introduced in (16) agrees with the one defined by Eq. (15).

When the homomorphism τ in Theorem 17.10 is an isomorphism, we can define the *index* of $g \in G$ with respect to the generators $g_1, \ldots, g_r$ as the unique element $\alpha = \langle k_1, \ldots, k_r \rangle$ of $\mathbf{Z}_{n_i} \times \cdots \times \mathbf{Z}_{n_r}$ for which $\tau(\alpha) = g$. Then the analogues of the remarks following Definition 17.1 are true. *Index* is the group isomorphism τ^{-1}; it allows us to perform a multiplication in G by adding indices. Each index is an r-tuple and the addition is modulo n_i in the ith place.

18. THE GROUP Φ(*p*)

This section is devoted to proving the following theorem and exploring its consequences.

18.1 Theorem. The group Φ(*p*) is cyclic if p is prime.

Proof. The theorem follows immediately from the fact that $\mathbf{Z}_p$ is a finite field (Theorem 9.5) and the theorem we are about to prove.

18.2 Theorem. A finite subgroup of the multiplicative group of a field is cyclic.

Proof. Let F be a field and n the order of a subgroup G of the multiplicative group F^* of nonzero elements of F. For each divisor d of n let $\Psi(d)$ be the number of elements of G of order d. Suppose $b \in G$ has order d. Then each of the d elements of the cyclic subgroup B generated by b is a root of the polynomial $x^d - 1$ in $F[x]$. Since that polynomial has at most d roots (Appendix 1, Corollary 18), we have found them all. Thus if $g \in G$ has order d it is in B and hence is one of the $\varphi(d)$ generator of B. Therefore for each $d \mid n$ either $\Psi(d) = 0$ or $\Psi(d) = \varphi(d)$. In either case $\Psi(d) \leq \varphi(d)$ so Theorem 17.6 implies G is cyclic.

Now it is time to translate some of the substance of this chapter into traditional number theoretic language.

18.3 Definition. The element $a \in \Phi(n)$ *belongs to the exponent d modulo n* if and only if $a^d \equiv 1(n)$, but $k < d$ implies $a^k \not\equiv 1(n)$. That is, a belongs to the exponent d means simply that d is the order of a in $\Phi(n)$. Thus 3 belongs to the exponent 2 modulo 8; Corollary 16.2 and Eqs. (1) and (3) show 10 belongs to the exponent 6 modulo both 7 and 13. An element $g \in \Phi(n)$ is a *primitive root* for n if and only if g belongs to the exponent $\varphi(n)$ modulo n. That is, g is a primitive root for n if and only if $\Phi(n)$ is cyclic and g is a generator. Corollary 16.2 says that $\lambda(n) = \varphi(n)$ if and only if 10 is a primitive root for n.

Theorem 18.1 says that primes have primitive roots. Some other integers do too; we shall soon see precisely which. Before we reach that point in our exposition you should experiment with the integers $n \leq 24$ to see which have primitive roots.

Whenever n has a primitive root g, computation in $\Phi(n)$ can be simplified by the construction of a table of indices. For example, successive doubling and reduction modulo 13 yields the data

k	0	1	2	3	4	5	6	7	8	9	10	11	12
2^k mod 13	1	2	4	8	3	6	12	11	9	5	10	7	1

$$(-1) \ (-2) \ (-4) \ (-8) \ (-3) \ (-6)$$

Once we reach the entry

$$2^6 \equiv 12 \not\equiv 1 \ (13)$$

we know 2 is a primitive root for 13, for the exponent to which it belongs must divide $\varphi(13) = 12$ and is greater than 6. The last entry

$$2^{12} \equiv 1 \ (13)$$

merely confirms Fermat's theorem. Finally, the entry above x in the second line of the table is just $\text{ind}_2 x$. As we remarked following Definition 17.1, $\text{ind}_2 : \Phi(13) \to Z_{12}$ is a group isomorphism. Let us use that fact to find the other primitive roots for 13. The element x is a primitive root for 13 if and only if x generates $\Phi(13)$, or, equivalently, $\text{ind}_2 x$ generates Z_{12}. Corollary 17.3 shows that happens just when

$$(\text{ind}_2 x, 12) = 1$$

that is, when

$$\text{ind}_2 x = 1, 5, 7, \text{ or } 11$$

so that 2, 6, 11, and 7 are the primitive roots for 13.

Here are two more examples which demonstrate uses of the index calculus.

What is the period of the decimal expansion of $1/13$?

$$\lambda(13) = \text{order of } 10 \text{ in } \Phi(13)$$
$$= \text{order of } \text{ind}_2(10) = 10 \text{ in } \mathbf{Z}_{12}$$
$$= \frac{12}{(10, 12)} \quad \text{(Lemma 17.2)}$$
$$= 6,$$

which we already knew.

Solve

$$x^8 \equiv 3 \ (13). \tag{17}$$

$$x^8 = 3 \text{ in } \Phi(13) \Leftrightarrow 8 \, \text{ind}_2 \, x = \text{ind}_2 3 \text{ in } \mathbf{Z}_{12}$$
$$\Leftrightarrow 8 \, \text{ind}_2 \, x \equiv 4 \ (12)$$
$$\Leftrightarrow \text{ind}_2 \, x \equiv 2 + 3n \ (12) \qquad \text{(Theorem 7.1)}$$
$$\Leftrightarrow \text{ind}_2 \, x = 2, 5, 8 \text{ or } 11$$
$$\Leftrightarrow \qquad x = 4, 6, 9, \text{ or } 7.$$

We showed above that 6 and 7 are primitive roots for 13. Since 6 and 7 each solve (17), we know

$$\text{ind}_6 3 = \text{ind}_7 3 = 8.$$

That equality is a coincidence; $\text{ind}_2 3 = 4 \neq 8$. In general the index of x in $\Phi(n)$ depends on the existence and the choice of a primitive root for n. So far we know of their existence only for primes.

The index calculus is only useful once a primitive root has been found; we have given no procedure other than trial and error for finding one. No universal shortcut is known though we shall show that in some special cases it is possible to find a primitive root for n without doing as much arithmetic as we required to find 2 for 13. When we locate a primitive root g by trial and error, the computations which prove g a primitive root also serve to build the table of indices to the base g. Appendix 2 contains a short table of primitive roots for primes.

The index calculus is of practical value if we have many actual computations to make modulo a fixed prime p. For most theoretical purposes what is important is just the existence of a primitive root. That is, we are often interested in consequences of the fact that $\Phi(p)$ is cyclic. For example,

Theorem 17.5 shows that when $d \mid p - 1$, $\Phi(p)$ contains $\varphi(d)$ elements which belong to the exponent d.   In particular, p has $\varphi(p - 1)$ primitive roots.

19.   THE GROUP $\Phi(2^{\alpha})$

$\Phi(2)$ has just one element and $\Phi(4)$ has $\varphi(4) = 2$, so each of these groups is cyclic.   However since

$$1^2 \equiv 3^2 \equiv 5^2 \equiv 7^2 \equiv 1 \ (8) \tag{18}$$

(see Problem 11.1), each of the four elements of $\Phi(8)$ is of order 2, so $\Phi(8)$ is not cyclic.   The group $\Phi(16)$ has 8 elements; 5 belongs to the exponent 4 modulo 16, but a little computation shows no element belongs to the exponent 8, so 16 has no primitive root.   This behavior persists.

Just the odd integers between 1 and 2^{α} are relatively prime to 2^{α}, so $\varphi(2^{\alpha}) = 2^{\alpha-1}$.   Every element of $\Phi(2^{\alpha})$ thus has order 2^{β} for some $\beta \leq \alpha - 1$.   The integer 2^{α} has a primitive root if and only if some element has order $2^{\alpha-1}$. We prove next that no such element exists when $\alpha \geq 3$.

19.1   Theorem.   If a is odd and $\alpha \geq 3$ then $a^{2^{\alpha-2}} \equiv 1(2^{\alpha})$.

Proof.   The congruences in (18) prove the theorem for $\alpha = 3$.   We proceed by induction.   Suppose the theorem true for α.   Then

$$a^{2^{\alpha-2}} = 1 + 2^{\alpha}t$$

for some integer t.   Squaring,

$$a^{2^{\alpha-1}} = 1 + 2\,(2^{\alpha}t) + 2^{2\alpha}t^2$$
$$\equiv 1\,(2^{\alpha+1}).$$

Here is another proof of Theorem 19.1 which is longer but more illuminating.

19.2   Lemma.   Let $\tau: G \to H$ be a group homomorphism.   Suppose $g \in G$ has order m and $\tau(g)$ has order n in H.   Let k be the order of the kernel K of τ.   Then

$$n \mid m \mid nk.$$

The proof is an exercise for the reader (Problem 22.11).

Theorem 19.1 is equivalent to the assertion that every element of $\Phi(2^{\alpha})$ has order $\leq 2^{\alpha-2}$ for $\alpha \geq 3$.   To prove that statement, reason as follows.

The natural ring homomorphism $\tau: \mathbf{Z}_{2_{\alpha}} \to \mathbf{Z}_8$ (see Problem 11.5) restricted to $\Phi(2^{\alpha})$ yields a group homomorphism of $\Phi(2^{\alpha})$ onto $\Phi(8)$. The kernel of this homomorphism has $\varphi(2^{\alpha})/\varphi(8) = 2^{\alpha-3}$ elements. Since every element of $\Phi(8)$ has order 2, Lemma 19.2 implies every element of $\Phi(2^{\alpha})$ has order $\leq 2 \cdot 2^{\alpha-3} = 2^{\alpha-2}$.

Thus the closest we can hope to come toward making $\Phi(2^{\alpha})$ cyclic when $\alpha \geq 3$ is to find an element of order $2^{\alpha-2}$, which would generate a cyclic subgroup of index 2. Fortunately, such an element is easy to find.

19.3 Theorem. The integer 5 belongs to the exponent $2^{\alpha-2}$ modulo 2^{α} when $\alpha \geq 3$.

Proof. Theorem 19.1 shows $5^{2^{\alpha-2}} \equiv 1(2^{\alpha})$. We need only show

$$5^{2^{\alpha-3}} \not\equiv 1 \ (2^{\alpha}). \tag{19}$$

To do this it is convenient to prove a little more, namely,

$$5^{2^{\alpha-3}} \equiv 1 + 2^{\alpha-1} \ (2^{\alpha}). \tag{20}$$

Since $1 + 2^{\alpha-1} \not\equiv 1(2^{\alpha})$, congruence (20) implies congruence (19). When $\alpha = 3$, (20) says simply that

$$5 \equiv 1 + 4 \ (8)$$

which is true. Suppose (20) true for α. Then

$$5^{2^{\alpha-3}} = 1 + 2^{\alpha-1} + 2^{\alpha}t$$

for some integer t. Square this trinomial:

$$5^{2^{\alpha-2}} = 1 + 2^{2\alpha-2} + 2^{2\alpha}t^2 + 2 \ (2^{\alpha-1} + 2^{\alpha}t + 2^{\alpha-1}2^{\alpha}t)$$

$$= 1 + 2^{\alpha} + 2^{\alpha+1}v \qquad \text{for some integer } v \text{ whose exact value is irrelevant}$$

$$\equiv 1 + 2^{\alpha} \ (2^{\alpha+1}).$$

Now we need only discover what is left over in $\Phi(2^{\alpha})$ outside the cyclic subgroup G_2 generated by 5. Since the order of G_2 is just half that of $\Phi(2^{\alpha})$, G_2 is of index 2 in $\Phi(2^{\alpha})$. Once we find a single element $b \notin G_2$, we can write $\Phi(2^{\alpha})$ as the disjoint union of the cosets G_2 and bG_2.

19.4 Lemma. $-1 \notin G_2$.

Proof. $5 \equiv 1(4)$ so $5^\rho \equiv 1(4)$ for all $\rho > 0$. Then $5^\rho \not\equiv -1(4)$ and hence

$$5^\rho \not\equiv -1 \ (2^\alpha)$$

for all $\rho > 0$ and $\alpha \geq 3$. But that says -1 is not congruent modulo 2^α to a power of 5. That is, $-1 \notin G_2$.

We have done more than to find an element outside G_2. The one we have generates a subgroup (with two elements) which meets G_2 only at 0, so that we can prove Theorem 19.5.

19.5   Theorem. $\Phi(2^\alpha)$ is isomorphic to $Z_2 \times Z_{2^{\alpha-2}}$ when $\alpha \geq 3$.

Proof. Apply Theorem 17.10 with $g_1 = -1$, $g_2 = 5$, $n_1 = 2$, and $n_2 = 2^{\alpha-2}$. The map $\tau: Z_2 \times Z_{2^{\alpha-2}} \to \Phi(2^\alpha)$ so constructed is surjective because -1 and 5 generate $\Phi(2^\alpha)$. Since both the domain and range of τ have $\varphi(2^\alpha) = 2^{\alpha-1}$ elements, τ is injective as well and hence is an isomorphism.

20.   THE GROUP $\Phi(p^\alpha)$

In this section we use techniques similar to those we just developed to study $\Phi(p^\alpha)$ when p is an odd prime. The results are nicer than those in Section 19. We shall discover that $\Phi(p^\alpha)$ is always cyclic by finding an integer g which is a primitive root for p^α for all α; g plays a role for p analogous to that played by 5 for 2.

20.1   Theorem.   Let g be a primitive root for p^α. Then just one of the following is true:

(a)   The integer g is a primitive root for $p^{\alpha+1}$;

(b) $g^{\varphi(p^\alpha)} \equiv 1 \ (p^{\alpha+1})$.

Proof. Let m be the exponent to which g belongs modulo $p^{\alpha+1}$. Then $m \mid \varphi(p^{\alpha+1})$. Moreover

$$g^m \equiv 1 \ (p^{\alpha+1})$$

so

$$g^m \equiv 1 \ (p^\alpha).$$

Hence

$$\varphi \ (p^\alpha) \mid m$$

since g is a primitive root for p^α. Therefore

$$\varphi\,(p^\alpha)|m|\varphi\,(p^{\alpha+1}) = p\varphi\,(p^\alpha) \tag{21}$$

which we could have proved directly by applying Lemma 19.2. Thus either

(a) $\quad m = \varphi\,(p^{\alpha+1}) \quad$ or $\quad$ (b) $\quad m = \varphi\,(p^\alpha)$

since p is prime.

Suppose g is a primitive root for p. Then $g^{p-1} \equiv 1(p)$, so $g^{p-1} = 1 + mp$ for some integer m. If $p \nmid m$, then Theorem 20.1 tells us g is a primitive root for p^2 as well. In fact, more will be true. We can prove the following analogue of Theorem 19.3.

20.2 Theorem. Let g be a primitive root for p such that

$$g^{p-1} = 1 + mp$$

and $p \nmid m$. Then g is a primitive root for p^α for all $\alpha > 0$.

Proof. Let the induction hypothesis be

$$g^{\varphi(p^\alpha)} = 1 + m_\alpha\,p^\alpha, \quad p \nmid m_\alpha, \tag{22}$$

which is true by assumption for $\alpha = 1$. Suppose it true for an $\alpha \geq 1$. Raise Eq. (22) to the pth power:

$$g^{\phi(p^{\alpha+1})} = g^{p\phi(p)}$$
$$= (1 + m_\alpha\,p)^p$$
$$= 1 + m_\alpha\,p^{\alpha+1} + tp^{2\alpha}$$
$$= 1 + m_{\alpha+1}p^{\alpha+1}$$

where t is an integer computed by collecting powers of m_α, binomial co-efficients, and powers of p, and

$$m_{\alpha+1} = m_\alpha + p^{2\alpha-1}t.$$

Since $p \nmid m_\alpha, p \nmid m_{\alpha+1}$, and (22) is true for all α. In particular, we now know

$$g^{\phi(p^\alpha)} \not\equiv 1\ (p^{\alpha+1})$$

for all α. Since g is a primitive root for p, repeated application of Theorem 20.1 yields alternative (a) each time.

Now all we need do is produce a g which satisfies the hypotheses of Theorem 20.2. For example, 2 is a primitive root for 3 and

$$2^{3-1} = 1 + 1 \cdot 3, \qquad 3 \nmid 1$$

so 2 will do for 3. However -1 is also a primitive root for 3 but clearly is not one for 9. Similarly, 2 is a primitive root for all powers of 5 since it is one for 5 and

$$2^{5-1} = 16 = 1 + 3 \cdot 5, \qquad 5 \nmid 3.$$

The integer 7 is also a primitive root for 5, but

$$7^4 = 49^2 \equiv (-1)^2 = 1 \ (25).$$

Alternative (b) in Theorem 20.1 occurs. Thus 7 is not a primitive root for 25.

20.3 Lemma. There is a primitive root g for p such that

$$g^{p-1} = 1 + pm \tag{23}$$

and

$$p \nmid m.$$

Proof. Let g be any primitive root for p. If $p \nmid m$ in (23), we are done. If we are unlucky and $m = kp$ let $g' = g + p \equiv g(p)$. Then g' is also a primitive root for p, and

$$
\begin{aligned}
(g')^{p-1} &= (g + p)^{p-1} \\
&= g^{p-1} + (p-1)pg^{p-2} + kp^2 \quad \text{for some integer } k \\
&= 1 + m'p
\end{aligned}
$$

where

$$m' = p(g^{p-2} + k) - g^{p-2}.$$

Since g is a primitive root for p, $(p, g) = 1$ so $p \nmid g^{p-2}$. Hence $p \nmid m'$, and g' satisfies the requirements of the lemma.

20.4 Theorem. Let p be an odd prime. Then $\Phi(p^\alpha)$ and $\Phi(2p^\alpha)$ are cyclic for all $\alpha > 0$.

Proof. We have just found a primitive root g for p^α, so $\Phi(p^\alpha)$ is cyclic. Moreover, we may assume g is odd, for if it is even then $g + p^\alpha$ is odd and is still a primitive root for p^α.

Then $(g, 2p^\alpha) = 1$. The exponent n to which g belongs modulo $2p^\alpha$ can be no less than that to which it belongs modulo p^α, which is $\varphi(p^\alpha)$. But

$$\varphi(2p^\alpha) = \varphi(2)\varphi(p^\alpha) = \varphi(p^\alpha)$$

so $n = \varphi(2p^\alpha)$, and g is a primitive root for $\Phi(2p^\alpha)$.

In the next section we shall show that we have found all the integers n for which $\Phi(n)$ is cyclic, namely, $n = 2, 4, p^\alpha$, or $2p^\alpha$.

21. THE GROUP $\Phi(n)$

We are now ready to complete our study of $\Phi(n)$. We shall show that $\Phi(mn)$ is isomorphic to $\Phi(m) \times \Phi(n)$ when m and n are relatively prime. Then we can find out all about $\Phi(n)$ by factoring n as a product of powers of primes and using our knowledge of the structure of the groups $\Phi(p^\alpha)$.

The route we follow is straightforward and computational. In the next theorem we use the Chinese remainder theorem and Theorem 17.10 to write $\Phi(n)$ as a product of cyclic groups.

21.1 Theorem. Let $n = 2^\alpha p_1^{\alpha_1} \cdots p_r^{\alpha_r}$ be the factorization of n into products of primes. Then $\Phi(n)$ is isomorphic to

$$\Phi(2^\alpha) \times \Phi(p_1^{\alpha_1}) \times \cdots \times \Phi(p_r^{\alpha_r})$$

which is in turn isomorphic to the product of cyclic groups

$$K = \mathbf{Z}_2 \times \mathbf{Z}_{2^{\alpha-2}} \times \mathbf{Z}_{\phi(p_1^{\alpha_1})} \times \cdots \times \mathbf{Z}_{\phi(p_r^{\alpha_r})} \tag{24}$$

when $\alpha \geq 3$. When $\alpha = 0$ or 1, omit the first two factors in Eq. (24); when $\alpha = 2$, omit the second factor.

Proof. We shall prove explicity only the case $\alpha \geq 3$. When $\alpha = 0, 1,$ or 2, the argument is similar but simpler.

Begin by choosing a primitive root b_i for $p_i^{\alpha_i}$, $i = 1, \ldots, r$ (Theorem 20.4). Let $b_{00} = -1$ and $b_0 = 5$; these are the analogues of primitive roots for 2^α.

Let h_{00} simultaneously solve the $r + 1$ congruences

$$x \equiv -1 \ (2^\alpha) \tag{25}$$

$$x \equiv 1 \ (p_i^{\alpha_i}) \qquad i = 1, \ldots, r. \tag{26}$$

Let h_0 simultaneously solve

$$x \equiv 5 \ (2^\alpha) \tag{27}$$

$$x \equiv 1 \ (p_i^{\alpha_i}) \qquad i = 1, \ldots, r. \tag{28}$$

For each j between 1 and r let h_j simultaneously solve

$$x \equiv 1 \ (2^\alpha) \tag{29}$$

$$x \equiv b_j \ (p_j^{\alpha_j}) \tag{30}$$

$$x \equiv 1 \ (p_i^{\alpha_i}) \qquad 1 \le i \le r, i \ne j. \tag{31}$$

The Chinese remainder theorem (8.1) guarantees the existence of h_{00}, $h_0, h_1, \ldots, h_r$. Finally let

$$g_i = \overline{h_i} \ \text{ in } \ \mathbf{Z}_n, \qquad i = 00, 0, 1, \ldots, r.$$

The order of g_i in $\Phi(n)$ is what it should be, namely, 2 when $i = 00$, $2^{\alpha-2}$ when $i = 0$ and $\varphi(p_i^{\alpha_i})$ when $i = 1, \ldots, r$. Let us see why this is true for g_1; the other cases are left to the reader. Let γ be the exponent to which g_1 belongs modulo n. Then

$$g_1^{\gamma} \equiv 1 \ (n)$$

implies

$$g_1^{\gamma} \equiv 1 \ (p_1^{\alpha_1})$$

so

$$\varphi \ (p_1^{\alpha_1}) \mid \gamma$$

because g_1 is a primitive root for $p_1^{\alpha_1}$. But the congruences (29), (30), and (31) imply $h_1^{\varphi(p_1\alpha_1)} \equiv 1$ modulo 2^α, $p_1^{\alpha_1}, \ldots, p_r^{\alpha_r}$ so

$$h_1^{\phi(p_1\alpha_1)} \equiv 1 \ (n).$$

Therefore

$$\gamma \mid \phi \ (p_1^{\alpha_1}).$$

Now we are ready to apply Theorem 17.10. Let $\tau: K \to \Phi(n)$ be the homomorphism constructed by that theorem from the data $g_{00}, g_0, \ldots, g_r$ given above. We wish to show τ is an isomorphism. Since K and $\Phi(n)$ each have order $\varphi(n)$, it will suffice to show that τ is injective.

To that end suppose $\beta = \langle \beta_{00}, \beta_0, \ldots, \beta_r \rangle$ is in the kernel of τ. We must show $\beta_{00} = \beta_0 = \cdots = \beta_r = 0$, where each equality is read in the appropriate cyclic factor of K. The typical "independence" argument works. We have assumed $\tau(\beta) \equiv 1(n)$, so

$$g_{00}^{\beta_{00}} g_0^{\beta_0} \cdots g_r^{\beta_r} = 1 + tn \tag{32}$$

for some integer t.

Our choice of the h_j and subsequent definition of the g_j shows that reducing Eq. (32) modulo $p_i^{\alpha_i}$ yields

$$b_i^{\beta_i} \equiv 1 \ (p_i^{\alpha_i}). \tag{33}$$

Since b_i is a primitive root for $p_i^{\alpha_i}$, (32) implies $\beta_i = 0$ in $\mathbf{Z}_{\varphi(p_i^{\alpha_i})}$. This argument works for each i between 1 and r, so we have reduced our problem to showing $\beta_{00} = \beta_0 = 0$ if

$$g_{00}^{\beta_{00}} g_0^{\beta_0} \equiv 1 \ (n). \tag{34}$$

Reduce Eq. (34) modulo 2^α:

$$(-1)^{\beta_{00}} 5^{\beta_0} \equiv 1 \ (2^\alpha)$$

which implies $\beta_{-1} = 0$ in $\mathbf{Z}_2$, and $\beta_0 = 0$ in $\mathbf{Z}_2{}^{\alpha-2}$ (Theorem 19.5).

The theorem is proved. We have made explicit the idea suggested in Example 8.3: To study congruences modulo n study first the corresponding congruences modulo the prime power divisors of n. In Appendix 3 we demonstrate that Theorem 21.1 is constructive by using it to answer a family of questions about $\Phi(315)$. In Problems 22.20 through 22.23 the reader is invited to apply Theorem 21.1 to develop the theory of the congruence

$$x^\beta \equiv m \ (n). \tag{35}$$

We now close this section and chapter with some corollaries to Theorem 21.1.

21.2   Corollary.   The function Φ is multiplicative.   That is, $\Phi(mn)$ is isomorphic to $\Phi(m) \times \Phi(n)$ when $(m, n) = 1$.

Proof.   The hypothesis $(m, n) = 1$ means that no prime which divides m divides n.   Thus to prove the corollary we need only group the cyclic factors of $\Phi(mn)$ provided by Theorem 21.1.

21.3   Definition.   The *universal exponent* $v(n)$ of n is the maximum of the orders of the elements of $\Phi(n)$.   Thus $v(n)/|\varphi(n)$, and $v(n) = \varphi(n)$ if and only if $\Phi(n)$ is cyclic.

In Section 19 we discovered that $v(4) = \varphi(4) = 2$ while $v(2^{\alpha}) = 2^{\alpha - 2} = \varphi(2^{\alpha})/2$ if $\alpha \geq 3$.   Since $\Phi(2)$ is the one element group, $v(2) = 1$.

21.4   Theorem.   Let $n = 2^{\alpha} p_1^{\alpha_1} \cdots p_r^{\alpha_r}$.   Then

$$v(n) = \text{l.c.m.}\{v(2^{\alpha}), \varphi(p_1^{\alpha_1}), \ldots, \varphi(p_r^{\alpha_r})\}. \tag{36}$$

Proof.   Apply Lemma 17.7 to the element of $\Phi(n)$ whose index is $\langle 1, 1, \ldots, 1 \rangle$ with respect to the isomorphism τ constructed in Theorem 21.1.   For example,

$$\begin{aligned}
\varphi(240) &= \varphi(2^4 \cdot 3 \cdot 5) \\
&= \varphi(2^4) \cdot \varphi(3) \cdot \varphi(5) \\
&= 8 \cdot 2 \cdot 4 \\
&= 64
\end{aligned}$$

so $\Phi(240)$ has 64 elements.   However

$$\begin{aligned}
v(240) &= \text{l.c.m.}\{v(2^4), \varphi(3), \varphi(5)\} \\
&= \text{l.c.m.}\{4, 2, 4\} \\
&= 4
\end{aligned}$$

so no element of $\Phi(240)$ has order greater than 4.   Since 64 is a power of 2 so is the order of every element of $\Phi(240)$.   Therefore

$$a^4 = a^{v(240)} \equiv 1 \ (240)$$

whenever $(a, 240) = 1$.   That is much stronger than Euler's theorem, which implies only

$$a^{64} \equiv 1 \ (240).$$

This suggested generalization of Euler's theorem is true even when $\varphi(n)$ and $v(n)$ are not powers of 2. The proof is Problem 22.17.

21.5 Corollary. If $(m, n) = 1$, then $v(mn) = \text{l.c.m.}\{v(m), v(n)\}$.

The proof is easy, so we omit it.

21.6 Theorem. The group $\Phi(n)$ is cyclic if and only if $n = 2, 4, p^\alpha$, or $2p^\alpha$ for an odd prime p.

Proof. We know that $\Phi(n)$ is cyclic for those n (Thorem 20.4 and the introductory paragraphs of Section 19). Suppose n is not one of those numbers. If n is a power of 2, then we showed in Section 19 that $\Phi(n)$ is not cyclic, so suppose n is divisible either by two odd primes p and q or by 4 and p. Consider the first alternative. Suppose p^α and q^β are the exact powers of p and q which divide n. Then $\Phi(p^\alpha q^\beta)$ is a factor of $\Phi(n)$ (Corollary 21.2). We shall show $\Phi(p^\alpha q^\beta)$ is not cyclic; Theorem 17.9 will then imply that $\Phi(n)$ is not cyclic.

Since p and q are odd, $\varphi(p^\alpha) = p^\alpha - p^{\alpha-1}$ and $\varphi(q^\beta)$ are both even. Therefore

$$v(p^\alpha q^\beta) = \text{l.c.m.}\{\varphi(p^\alpha), \varphi(q^\beta)\}$$
$$\leq \frac{\varphi(p^\alpha)\varphi(q^\beta)}{2}$$
$$< \varphi(p^\alpha)\varphi(q^\beta)$$
$$= \varphi(p^\alpha q^\beta).$$

Thus $\Phi(p^\alpha q^\beta)$ is not cyclic.

A similar argument covers the second alternative. If $8 \,|\, n$, then $\Phi(n)$ has the noncyclic factor $\Phi(2^\alpha)$ and hence is not cyclic. If $4 \,|\, n$ but 8 does not, then

$$v(4p^\alpha) = \text{l.c.m.}\{2, \varphi(p^\alpha)\}$$
$$= \varphi(p^\alpha)$$
$$< 2\varphi(p^\alpha)$$
$$= \varphi(4)\varphi(p^\alpha) = \varphi(4p^\alpha)$$

implies $\Phi(n)$ has the noncyclic factor $\Phi(4p^\alpha)$.

22. PROBLEMS

22.1 Divisibility tests again. Let $m, a_0, a_1, \ldots, a_k$ be as in Problem 11.12 and $n, r_1, r_2, \ldots$, as in Section 16. Prove

$$m \equiv a_0 + r_1 a_1 + \cdots + r_k a_k \ (n). \tag{37}$$

Show that Problem 11.12 consists of special cases of (37). How does this problem solve Problem 11.13?

22.2 Show that the usual "long division algorithm" does in fact yield the decimal expansion of $1/n$.

22.3* In the expansion of $1/7$ given by Eq. (1) the sum of the digits in places i and $i + 3$ is always nine: $1 + 8 = 4 + 5 = 2 + 7 = 9$. Call n a *nines number* when $(10, n) = 1$ and the digits in places i and $(\lambda(n)/2) + i$ in its decimal expansion always sum to 9.

(a) A necessary condition that n be a nines number is that $\lambda(n)$ be even. Does that condition suffice?

(b) Show that n is a nines number if and only if $10^{\lambda(n)/2} \equiv -1(n)$.

(c) Prove. If p is prime, then $\lambda(p)$ is even if and only if p is a nines number.

You will find these and related results in Leavitt, W. G., "A Theorem on Repeating Decimals," *Amer. Math. Monthly*, **74**, (1967), 669–673.

22.4 Let g and h both generate the cyclic group G. Derive a formula (analogous to the one you learned in secondary school relating logarithms to different bases) with which you can compute $\text{ind}_g(x)$ from $\text{ind}_h(x)$.

22.5 Prove that every subgroup of a cyclic group is cyclic. Use that fact to replace the part of the proof of Corollary 17.5 which depends on Eq. (10) of Chapter 2. Then deduce that equation from Corollary 17.5.

22.6 Let g and h be commuting elements of the group G. Show that the order of gh divides the least common multiple m of the orders of g and h. Need it equal m? (Compare Lemma 17.7.) Write $n = p_1^{\alpha_1} \cdots p_k^{\alpha_k}$ as a product of powers of distinct primes. Show that the least m for which S_m (the group of permutations of m symbols) contains an element of order n is $m = p_1^{\alpha_1} + \cdots + p_k^{\alpha_k}$.

22.7 Deduce Theorem 10.4 from Corollaries 17.5 and 17.8.

22.8 Prove: If a belongs to the exponent 3 modulo the prime p, then

$$1 + a + a^2 \equiv 0 \ (p).$$

Generalize. Can you do without the assumption that p is prime?

22.9 Find a primitive root for 17; prepare a table of indices and use it to solve the congruences

$$x^{20} \equiv 13 \ (17)$$
$$x^{12} \equiv 13 \ (17)$$
$$x^{48} \equiv 9 \ (17)$$
$$x^{11} \equiv 9 \ (17).$$

What are the periods of the decimal and binary expansions of $1/17$?

22.10 Let p be an odd prime, and suppose $p \nmid a$. Show that

$$x^2 \equiv a \ (p^\alpha)$$

has a solution if and only if

$$x^2 \equiv a \ (p)$$

has and that when solutions exist there are two of them modulo p^α.

22.11 Prove Lemma 19.2.

22.12 Let a be an odd integer and $\alpha \geq 3$. Use the knowledge of the structure of $\Phi(2^\alpha)$ set out in Section 16 to show that

$$x^2 \equiv a \ (2^\alpha)$$

can be solved if and only if

$$a \equiv 1 \ (8)$$

and that when solutions exist there are four of them modulo 2^α.

22.13* Work out a new proof of Theorem 20.3 along the following lines. The kernel K of the natural map

$$\Phi \ (p^{\alpha+1}) \rightarrow \Phi \ (p^\alpha)$$

contains the elements of $\Phi(p^{\alpha+1})$ which are congruent to 1 modulo p^α. Show K is naturally isomorphic to the cyclic group Z_p, $+$. Then use the fact that $\Phi(p^{\alpha+1})/K$ is isomorphic to $\Phi(p^\alpha)$, which, inductively, you may assume is cyclic.

22.14 Let g be a primitive root for $n > 2$. Prove

$$\text{ind}_g \ (-1) = \frac{\varphi(n)}{2}$$

and deduce

$$\text{ind}_g \ (-a) \equiv \text{ind}_g \ (a) + \frac{\varphi(n)}{2} \ (\varphi(n)).$$

Thus only half a table of indices need be prepared with brute force. This problem explains the presence of the parenthetical negative integers beneath the second half of the table in Section 18.

22.15 Let g be a primitive root for the odd prime p. When will $-g$ be one too?

22.16 Show that the product of the primitive roots of the prime $p > 3$ is congruent to 1 modulo p.

22.17 Prove $a^{v(n)} \equiv 1(n)$ when $(a, n) = 1$.   (See the example following Theorem 21.4.)

22.18 Prove that 3 belongs to the universal exponent $v(2^\alpha) = 2^{\alpha-2}$ modulo 2^α for every $\alpha \geq 3$.

22.19 Call n an F-number if Fermat's theorem (9.7) is true for n, that is, if

$$a^{n-1} \equiv 1 \ (n)$$

whenever $(a, n) = 1$.

Show 561 is an F-number.   It happens to be the smallest composite F-number; The next two are 1105 and 1729.   For more information on F-numbers see Ore, Chapter 14 (Reference 6 in the Bibliography).

In the next problems we build a theory for solving the congruence

$$x^\beta \equiv m \ (n), \qquad (m, n) = 1. \tag{38}$$

The method is to establish some easy facts about products of groups and then to apply Theorem 21.1.   The results generalize parts of Problems 22.10 and 22.12.

22.20 Let β be an integer and

$$\sigma_\beta \ (x) = x^\beta.$$

This definition makes sense whenever x belongs to a group.   When G is an abelian group $\sigma_\beta : G \to G$ is a homomorphism (why?); then let G_β be its kernel.   Prove
  (a) $(G \times H)_\beta = G_\beta \times H_\beta$.
  (b)   Given $g \in G$ the solutions (if any) in G to

$$x^\beta = g$$

form a coset of G_β.

22.21 Let $k_\beta(n)$ be the number of solutions in $\Phi(n)$ to the congruence

$$x^\beta \equiv 1 \ (n).$$

(a)   Prove k_β is multiplicative.
(b)   Show $k_\beta(p^\alpha) = (\beta, \varphi(p^\alpha))$ when p is an odd prime.
(c)   What is $k_\beta(2^\alpha)$?

22.22 (a)   Show (38) can be solved for $\varphi(n)/k_\beta(n)$ elements m of $\Phi(n)$.
(b)   When (38) has a solution, it has exactly $k_\beta(n)$ of them in $\Phi(n)$.
(c)   If $\beta \equiv \beta' \ (v(n))$, then (38) is equivalent to

$$x^{\beta'} \equiv m \ (n).$$

(d)   Congruence (38) has a solution for every m if and only if $(\beta, v(n)) = 1$.

22.23 Solve

$$x^8 \equiv 256 \ (315)$$
$$x^8 \equiv 263 \ (315)$$
$$x^5 \equiv 256 \ (315)$$
$$x^5 \equiv 263 \ (315)$$

using either the techniques of the problems above or those developed in Appendix 3.

22.24* Show that

$$x^4 \equiv -1 \ (p)$$

has a solution for the odd prime p if and only if $p \equiv 1(8)$. Deduce that there are infinitely many primes of this kind.

22.25* When does

$$x^{2^\alpha} \equiv -1 \ (p)$$

have a solution? (Compare Problem 22.4 and Theorem 13.2.)

5

Quadratic Reciprocity

In this chapter we shall discuss the congruence

$$x^2 \equiv a \ (p) \tag{1}$$

where p is a prime which does not divide a.

We solved this problem in principle in Section 18; choose a primitive root g for p and solve

$$2 \operatorname{ind}_g (x) \equiv \operatorname{ind}_g (a) (p - 1). \tag{2}$$

Unfortunately, the computation of indices is nontrivial. Our objective now is Gauss's Quadratic Reciprocity Law, with which we can decide quite easily whether or not (1) has a solution. Finding one still requires the index calculus or some equivalent.

We have already seen and solved one special case of (1), namely,

$$x^2 \equiv -1 \ (p)$$

has a solution if and only if $p \equiv 1(4)$ (Theorem 13.2). This fact allowed us to prove that there are infinitely many primes congruent to 1 modulo 4.

The Quadratic Reciprocity Law will help us show that some other arithmetic progressions contain infinitely many primes.   We shall also use it to study the Diophantine equation

$$x^2 - my^2 = p$$

which we looked at in Chapter 3 for $m = 1$ and $m = -1$.

23.   RESIDUES

We say that a is an mth *power residue* of n when $(a, n) = 1$ and the congruence

$$x^m \equiv a \ (n) \tag{3}$$

has a solution.   Problems 22.9, 22.10, 22.12, and 22.20 through 22.25 take up various aspects of the theory of mth power residues.   We are interested now in the special case $m = 2$, n prime, when (3) coincides with (1).   We say then that a is a *quadratic residue*, or *residue*, of p.   Numbers which are not residues are *nonresidues*.   Since every odd integer a is an mth power residue of 2 for every m, we shall restrict our attention to odd primes p.   The introductory remarks above show that -1 is a residue of p if and only if p is congruent to 1 modulo 4.

We shall follow our customary useful but ambiguous procedure and think of the residues of p both as integers and as elements of $\mathbf{Z}_p$.   We can find the residues of p by squaring each of the elements of $\Phi(p)$.   For example,

$$(\pm 1)^2 \equiv 1 \ (7)$$
$$(\pm 2)^2 \equiv 4 \ (7)$$

and

$$(\pm 3)^2 \equiv 2 \ (7)$$

so 1, 2, and 4 are the residues of 7.

23.1   Lemma.   Let g be a primitive root for p.   Then $a \in \Phi(p)$ is a residue of p if and only if $\text{ind}_g(a)$ is even.

Proof.   Congruence (2) can be solved for $\text{ind}_g(x) \in \mathbf{Z}_{p-1}$ if and only if $(2, p - 1) = 2 \mid \text{ind}_g(a)$ (Theorem 7.1).

23.2   Theorem.   The set R of residues of p is a subgroup of $\Phi(p)$ of order $\varphi(p)/2 = (p - 1)/2$.

Proof. The set R is a subgroup since it is the image of the homomorphism $x \rightsquigarrow x^2$ mapping $\Phi(p)$ into itself. Since just half the elements of $\Phi(p)$ have even indices, R has order $(p-1)/2$.

It is customary to identify the two element group $\Phi(p)/R$ with the group $\{\pm 1\}$ under multiplication. Let

$$\tau: \Phi(p) \to \frac{\Phi(p)}{R} = \{\pm 1\} \tag{4}$$

be the natural map.

The classical and useful *Legendre symbol* is defined for integers a and odd primes p by

$$\left(\frac{a}{p}\right) = \begin{cases} \tau(\bar{a}) & \text{when } p \nmid a \\ 0 & \text{when } p \mid a \end{cases}$$

$$= \begin{cases} 1 & \text{when } a \text{ is a residue of } p \\ -1 & \text{when } a \text{ is not a residue of } p \\ 0 & \text{when } p \mid a. \end{cases}$$

When $p \nmid a$, think of $\left(\frac{a}{p}\right)$ as the answer to the question "Is a a residue of p?" The answer $+1$ means yes, -1, no.

23.3 Theorem.

(a) $\left(\frac{a}{p}\right) = \left(\frac{b}{p}\right)$ if $a \equiv b\ (p)$

(b) $\left(\frac{ab}{p}\right) = \left(\frac{a}{p}\right)\left(\frac{b}{p}\right)$

(c) $\left(\frac{a^2}{p}\right) = 1$ if $p \nmid a$

(d) $\left(\frac{1}{p}\right) = 1.$

Proof. Part (a) is obvious; (b) is clearly true when $p \mid ab$. If neither a nor b is a multiple of p, then (b) follows from the fact that reduction modulo p is a ring homomorphism from $\mathbf{Z}$ to $\mathbf{Z}_p$, and τ is a group homomorphism. Part (c) is a special case of (b) and (d) a special case of (c).

The results of our computations above of the residues of 7 may now be summarized as

$$\left(\frac{1}{7}\right) = \left(\frac{2}{7}\right) = \left(\frac{4}{7}\right) = 1$$

$$\left(\frac{3}{7}\right) = \left(\frac{5}{7}\right) = \left(\frac{6}{7}\right) = -1 \tag{5}$$

$$\left(\frac{0}{7}\right) = 0.$$

Our knowledge of the solvability of $x^2 \equiv -1(p)$ becomes

$$\left(\frac{-1}{p}\right) = (-1)^{(p-1)/2} \tag{6}$$

since the exponent in the right member of Eq. (6) is even if and only if $p \equiv 1(4)$. Eq. (6) is a special case of the following theorem due to Euler.

23.4 Theorem. $\left(\dfrac{a}{p}\right) \equiv a^{(p-1)/2} \ (p)$

Proof. Since both members of the desired congruence are 0 when $p \mid a$, we may restrict our attention to the case $p \nmid a$. Then Fermat's theorem implies

$$(a^{(p-1)/2})^2 = a^{p-1} \equiv 1 \ (p). \tag{7}$$

Since the polynomial $x^2 - 1 = (x - 1)(x + 1)$ has just two roots ± 1 in the field $\mathbf{Z}_p$, congruence (7) implies

$$a^{(p-1)/2} \equiv \pm 1 \ (p). \tag{8}$$

We need only decide which. When $a \equiv b^2(p)$ is one of the $(p - 1)/2$ residues of p,

$$a^{(p-1)/2} \equiv (b^2)^{(p-1)/2} \equiv b^{p-1} \equiv 1 \ (p)$$

so $+1$ occurs in (8). Thus the residues of p are the roots of the polynomial $x^{(p-1)/2} - 1$ in $\mathbf{Z}_p$. Since that polynomial has at most $(p - 1)/2$ roots, all the roots are accounted for, so -1 must occur in (8) when a is a nonresidue.

24. THE LEMMA OF GAUSS

In this section we count again in $\Phi(p)$ to find two more expressions for $\left(\dfrac{a}{p}\right)$.
We start by introducing the "greatest integer" function [   ].

24.1 Definition. The largest integer less than or equal to x is denoted by $[x]$. Thus

$$[2.1] = 2, \qquad [\pi] = 3,$$
$$[1] = 1, \qquad [-6.4] = -7.$$

If n is an integer, then the largest multiple of n which is less than or equal to x is $[x/n]n$, so that

$$x = \left[\frac{x}{n}\right]n + r, \qquad 0 \le r < n \tag{9}$$

for every real number x. When x is an integer, Eq. (9) shows that the function [   ] is useful for describing the quotient in the long division algorithm (Lemma 2.2). That is the property of [   ] we need now.

24.2 The Lemma of Gauss. Suppose a and the odd prime p are relatively prime. Consider the $(p-1)/2$ principal remainders r_j defined by

$$ja = \left[\frac{ja}{p}\right]p + r_j \tag{10}$$

for $j = 1, 2, \ldots, (p-1)/2$. Each r_j satisfies $0 < r_j < p$. Let n be the number of j for which $r_j > p/2$. Then

$$\left(\frac{a}{p}\right) = (-1)^n.$$

That is, a is a residue of p if and only if n is even.

Before we prove the lemma let us see what it says in two familiar cases and one unfamiliar one.
Suppose $a = 1$. Then

$$ja = j = 0 \cdot p + j$$

so

$$r_j = j, \qquad j = 1, \ldots, \frac{p-1}{2}.$$

No remainder is larger than $p/2$, $n = 0$, and

$$\left(\frac{1}{p}\right) = (-1)^0 = 1$$

which is nothing new.

Suppose $a = -1$. Then

$$ja = -j = -p + (p - j),$$

that is,

$$\left[\frac{-j}{p}\right] = -1 \quad \text{and} \quad r_j = p - j$$

for $j = 1, \ldots, (p-1)/2$. Thus every remainder is larger than $p/2$, $n = (p-1)/2$ and

$$\left(\frac{-1}{p}\right) = (-1)^{(p-1)/2}$$

which is Eq. (6) again.

Suppose $a = 2$. Then

$$ja = 2j = 0 \cdot p + 2j$$

so

$$r_j = 2j, \qquad j = 1, \ldots, \frac{p-1}{2}.$$

Now

$$r_j = 2j > \frac{p}{2}$$

if and only if

$$j > \frac{p}{4}.$$

Since there are $[p/4]$ positive integers less than $p/4$, there are

$$n = \frac{p-1}{2} - \left[\frac{p}{4}\right]$$

remainders r_j greater than $p/2$. How can we discover the parity of n in terms of p? The key is to write p modulo 8. Since p is odd

$$p = 8k + v$$

where $v = 1, 3, 5,$ or 7. Next compile the table

v	$\dfrac{(8k+v)-1}{2}$	$\left[\dfrac{8k+v}{4}\right]$	n
1	$4k$	$2k$	$2k$
3	$4k+1$	$2k$	$2k+1$
5	$4k+2$	$2k+1$	$2k+1$
7	$4k+3$	$2k+1$	$2k+2.$

Thus n is even if and only if $v = 1$ or 7. We have proved the next theorem.

24.3 Theorem. $\left(\dfrac{2}{p}\right) = 1$ if and only if $p \equiv \pm 1(8)$. For example,

$$\left(\frac{2}{17}\right) = \left(\frac{2}{23}\right) = 1 \quad \text{but} \quad \left(\frac{2}{19}\right) = -1.$$

24.4 Corollary. There are infinitely many primes congruent to ± 1 modulo 8.

Proof. There are infinitely many primes modulo which the polynomial $x^2 - 2$ has a root (Theorem 12.1); these are the primes for which 2 is a residue. (Compare Problem 22.24.)

Now we return to the general stituation and prove the lemma of Gauss. Let $s_1, \ldots, s_n$ be the remainders $r_j > p/2$. Let $m = ((p-1)/2) - n$ and $t_1, \ldots, t_m$ be the remainders $r_j < p/2$. Let

$$u_1 = p - s_1, \ldots, u_n = p - s_n.$$

Then

$$1 \leq u_j < \frac{p}{2} \quad j = 1, \ldots, n$$

and

$$1 \leq t_i < \frac{p}{2} \quad i = 1, \ldots, m = \frac{p-1}{2} - n. \tag{11}$$

Let us show that the $(p-1)/2$ integers

$$u_1, \ldots, u_n, \quad t_1, \ldots, t_m \tag{12}$$

are incongruent modulo p. Observe that for any choice of signs the $(p-1)/2$ numbers

$$\pm 1, \pm 2, \ldots, \pm \frac{p-1}{2}$$

are mutually incongruent modulo p. Then since $\Phi(p)$ is a group, the sequence

$$\pm a, \pm 2a, \ldots, \pm \frac{p-1}{2} a \tag{13}$$

also represents $(p-1)/2$ distinct elements of $\Phi(p)$ for any choice of signs. If $r_j < p/2$, then for some i

$$t_i = r_j \equiv ja \ (p)$$

while if $r_j > p/2$, then for some i

$$u_i = p - s_i = p - r_j \equiv -ja \ (p).$$

Therefore the sequence (12) of t_i's and u_i's is congruent modulo p to a rearrangement of the sequence (13) for a particular choice of n minus signs.

Hence the t_i's and u_i's are mutually incongruent modulo p. This fact and the inequalities in (11) together imply that the sequence (12) is just a re-arrangement of the sequence $1, 2, \ldots, (p-1)/2$. Therefore

$$\left(\frac{p-1}{2}\right)! = t_1 \cdots t_m u_1 \cdots u_n$$

$$\equiv (-1)^n \left(\frac{p-1}{2}\right)! \, a^{(p-1)/2} \, (p). \tag{14}$$

Cancel $\left(\dfrac{p-1}{2}\right)!$ in congruence (14). Then since $(-1)^n = (-1)^{-n}$,

$$a^{(p-1)/2} \equiv (-1)^n \, (p)$$

which together with Theorem 23.4 proves the lemma of Gauss.

24.5   Theorem. $\left(\dfrac{-2}{p}\right) = 1$ if and only if $p \equiv 1$ or 3 modulo 8; there are infinitely many primes of this kind.

Proof. We could compute this directly from the lemma of Gauss (see Problem 27.11) but prefer to use an algebraic shortcut.

$$\left(\frac{-2}{p}\right) = \left(\frac{-1}{p}\right)\left(\frac{2}{p}\right)$$

(Part (b) of Theorem 23.1), so $\left(\dfrac{-2}{p}\right) = 1$ if and only if

$$\left(\frac{-1}{p}\right) = \left(\frac{2}{p}\right) = \pm 1. \tag{15}$$

The right member of Eq. (15) is $+1$ when

$$p \equiv 1(4) \quad \text{and} \quad p \equiv \pm 1(8),$$

that is, when $p \equiv 1(8)$. The right member is -1 when

$$p \equiv 3(4) \quad \text{and} \quad p \equiv \pm 3(8),$$

that is, when $p \equiv 3(8)$.

The proof of the infinitude of primes congruent to 1 or 3 modulo 8 is just like the proof of Corollary 24.4.

Now we know all about $\left(\dfrac{\pm 2}{p}\right)$. Since $\left(\dfrac{ab}{p}\right) = \left(\dfrac{a}{p}\right)\left(\dfrac{b}{p}\right)$, we need only work at evaluating $\left(\dfrac{a}{p}\right)$ when a is odd.

24.6 Lemma. Let a be odd, $p \nmid a$, and

$$M = \left[\frac{a}{p}\right] + \left[\frac{2a}{p}\right] + \cdots + \left[\frac{((p-1)/2)a}{p}\right]. \tag{16}$$

Then

$$\left(\frac{a}{p}\right) = (-1)^M.$$

Proof. We shall keep the notation established in the proof of the lemma of Gauss. In light of that lemma we need only prove $M \equiv n(2)$, for then $(-1)^M = (-1)^n$. Let

$$T = 1 + 2 + \cdots + \frac{p-1}{2}.$$

(T happens to be $(p^2 - 1)/8$, but we shall not need its exact value.) Now sum Eq. (10) over j for $j = 1, \ldots, (p-1)/2$. Then

$$aT = pM + \sum_{j=1}^{(p-1)/2} r_j.$$

Thus modulo 2

$$
\begin{aligned}
T &\equiv aT &&(2) &&(a \text{ is odd}) \\
&\equiv pM + \sum r_j\,(2) \\
&\equiv M + \sum r_j\,(2) &&(p \text{ is odd}).
\end{aligned} \tag{17}
$$

We also know from our proof of the lemma of Gauss that

$$
\begin{aligned}
T &= \sum_{i=1}^{n} u_i + \sum_{i=1}^{m} t_i \\
&= np - \sum_{i=1}^{n} s_i + \sum_{i=1}^{m} t_i.
\end{aligned} \tag{18}
$$

Modulo 2 we may ignore the p and the minus sign in Eq. (18), so

$$T \equiv n + \sum s_i + \sum t_i \quad (2)$$
$$= n + \sum r_j. \quad (19)$$

Compare (17) with (19) to deduce

$$M \equiv n \ (2).$$

25. THE QUADRATIC RECIPROCITY LAW

We must keep the notation introduced in Section 24 a little while longer. A *lattice point* in the Cartesian plane $\mathbf{R} \times \mathbf{R}$ is a point both of whose coordinates are integers.

25.1 Lemma (Eisenstein). Suppose the odd prime p does not divide the odd integer a. Consider the right triangle $\triangle$ in the Cartesian plane with sides

$$x = \frac{p}{2},$$

$$y = 0 \quad \text{(the } x \text{ axis)},$$

and

$$y = \frac{a}{p} x.$$

Then there are no lattice points other than $\langle 0, 0 \rangle$ on the hypotenuse of $\triangle$ and exactly M lattice points in its interior.

Proof. Figure 1 may help. The hypotenuse cannot contain a lattice point other than the origin since if j is a positive integer less than $p/2$, then aj/p is not an integer.

Let us count the lattice points inside $\triangle$ by looking at each vertical line. Suppose j is an integer, and $1 \leq j \leq (p-1)/2 < p/2$. There are exactly $[aj/p]$ positive integers less than aj/p, so there are exactly $[aj/p]$ lattice points inside $\triangle$ on the line $x = j$. Then sum over j.

25.2 Theorem (Quadratic Reciprocity). If p and q are odd primes then

$$\left(\frac{p}{q}\right)\left(\frac{q}{p}\right) = (-1)^{((p-1)/2)((q-1)/2)}.$$

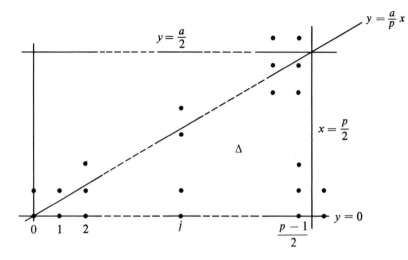

Figure 1

Proof. As usual, let

$$M = \sum_{j=1}^{(p-1)/2} \left[\frac{jq}{p}\right]$$

and define

$$N = \sum_{j=1}^{(q-1)/2} \left[\frac{jp}{q}\right].$$

Then Lemma 24.6 implies

$$\left(\frac{q}{p}\right) = (-1)^M \quad \text{and} \quad \left(\frac{p}{q}\right) = (-1)^N.$$

Thus it suffices to show

$$M + N \equiv \frac{p-1}{2} \cdot \frac{q-1}{2} \quad (2). \tag{20}$$

We shall in fact show more; the two members of (20) are equal.
    Consider the rectangle Γ with sides

$$y = 0, \quad y = \frac{q}{2}, \quad x = 0, \quad x = \frac{p}{2}.$$

The diagonal D of Γ has the equation

$$y = \frac{q}{p} x.$$

Lemma 25.1 shows D contains no lattice points and that there are M lattice points inside Γ under D.

Since q too is an odd prime, we may apply Lemma 25.1 with p and q in place of a and p. If we also interchange the roles of x and y, it follows that there are N lattice points inside Γ over D.

Since $[p/2] = (p-1)/2$, and $[q/2] = (q-1)/2$ there are

$$\frac{p-1}{2} \cdot \frac{q-1}{2} = M + N$$

lattice points inside R.

Theorem 25.2 is too important to be stated so starkly. Let us rephrase it. All that matters in the formula connecting the Legendre symbols is the parity of the exponent. The integer $(p-1)/2$ is even if and only if $p \equiv 1(4)$ so that the Law of Quadratic Reciprocity says that

$$\left(\frac{p}{q}\right)\left(\frac{q}{p}\right) = 1$$

unless *both* p and q are congruent to 3 modulo 4, in which case

$$\left(\frac{p}{q}\right)\left(\frac{q}{p}\right) = -1.$$

Restated again,

$$\left(\frac{p}{q}\right) = \left(\frac{q}{p}\right)$$

unless $p \equiv q \equiv 3(4)$, in which case

$$\left(\frac{p}{q}\right) = -\left(\frac{q}{p}\right).$$

We can use quadratic reciprocity to answer questions such as "Is 3 a residue of 41?" Both 3 and 41 are prime, and $3 \equiv 3(4)$ while $41 \equiv 1(4)$.

Therefore

$$\left(\frac{3}{41}\right) = \left(\frac{41}{3}\right)$$

Now apply Theorem 23.3: $41 \equiv 2(3)$ so

$$\left(\frac{41}{3}\right) = \left(\frac{2}{3}\right).$$

Finally,

$$\left(\frac{2}{3}\right) = -1$$

either because $3 \equiv 3 \not\equiv \pm 1(8)$ (Theorem 24.3) or, better, because by inspection

$$x^2 \equiv 2(3)$$

has no solution.   Therefore

$$\left(\frac{3}{41}\right) = -1$$

and 3 is not a residue of 41.

Here is a similar computation.   The reader should provide a reason for each step.

$$\left(\frac{31}{41}\right) = \left(\frac{-10}{41}\right) = \left(\frac{-1}{41}\right)\left(\frac{5}{41}\right)\left(\frac{2}{41}\right)$$

$$= 1 \cdot \left(\frac{41}{5}\right) \cdot 1 = \left(\frac{1}{5}\right) = 1$$

so the congruence

$$x^2 \equiv 31 \ (41)$$

has a solution.

For which primes is 3 a residue? Clearly 3 is a residue of 2. Suppose p is a prime other than 2 or 3. Since $3 \equiv 3(4)$,

$$\left(\frac{3}{p}\right) = \begin{cases} \left(\dfrac{p}{3}\right) & \text{if } p \equiv 1\ (4) \\ -\left(\dfrac{p}{3}\right) & \text{if } p \equiv 3\ (4). \end{cases}$$

We know $p \equiv 1$ or $2(3)$ and that

$$\left(\frac{1}{3}\right) = 1; \qquad \left(\frac{2}{3}\right) = -1$$

so

$$\left(\frac{3}{p}\right) = 1$$

if and only if

$$p \equiv 1\ (4) \quad \text{and} \quad p \equiv 1\ (3) \tag{21}$$

or

$$p \equiv 3\ (4) \quad \text{and} \quad p2 \equiv (3). \tag{22}$$

The conditions in (21) are equivalent to $p \equiv 1(12)$; those in (22) to $p \equiv -1(12)$. Thus we have proved the following theorem.

25.3 Theorem. The Legendre symbol $\left(\dfrac{3}{p}\right) = 1$ if and only if $p = 2$ or $p \equiv \pm 1(12)$; there are infinitely many primes of this kind.

25.4 Theorem. The Legendre symbol $\left(\dfrac{-3}{p}\right) = 1$ if and only if $p = 2$ or $p \equiv 1(6)$; there are infinitely many primes of this kind.

Proof. Theorem 25.4 is to Theorem 25.3 as 24.5 is to 24.3. The proof is as easy. Suppose $p \neq 2$. Then

$$\left(\frac{-3}{p}\right) = \left(\frac{-1}{p}\right)\left(\frac{3}{p}\right)$$

which is 1 if and only if

$$p \equiv 1 \ (4) \quad \text{and} \quad p \equiv \pm 1 \ (12) \tag{23}$$

or

$$p \equiv 3 \ (4) \quad \text{and} \quad p \equiv \pm 1 \ (12). \tag{24}$$

The conditions in (23) imply $p \equiv 1(12)$; those in (24) imply $p \equiv -5 \equiv 7(12)$. Both cases are covered exactly by $p \equiv 1(6)$.

26. THE DIOPHANTINE EQUATION $x^2 - my^2 = \pm p$

In this section we shall exploit the Law of Quadratic Reciprocity and Thue's theorem to study

$$x^2 - my^2 = \pm p \tag{25}$$

for a few small values of m. Some cases have been covered already. We discovered in Section 14 that

$$x^2 + y^2 = p$$

has a solution if and only if $p = 2$ or $p \equiv 1(4)$ or, equivalently, $\left(\dfrac{-1}{p}\right) = 1$. In the same section we also proved that

$$x^2 - y^2 = p$$

has a solution for every odd prime p. Of course $\left(\dfrac{1}{p}\right) = 1$ for every odd prime too.

These remarks should suggest a connection between $\left(\dfrac{m}{p}\right)$ and the existence of solutions to Eq. (25). Let us formalize and generalize the suggestion.

26.1 Theorem. If

$$x^2 - my^2 = kp \tag{26}$$

and p is an odd prime which divides neither k nor m, then $\left(\dfrac{m}{p}\right) = 1$.

Proof. First we show that $(p, x) = (p, y) = 1$, in analogy to Lemma 14.1. If p divides x or y, then Eq. (26) implies p divides x and y since $p \nmid m$. Thus

$$p^2 \mid x^2 - my^2 = kp$$

but $p \nmid k$, a contradiction. Now reduce Eq. (26) modulo p:

$$x^2 \equiv my^2 \ (p)$$

so

$$(xy^{-1})^2 \equiv m \ (p). \tag{27}$$

The formal fraction in (27) looks out of place and must be explained. Here y^{-1} is the inverse of y in $\Phi(p)$; we could write y^{p-2} instead (Euler's Theorem). Equation (27) says m is a residue of p, so

$$\left(\frac{m}{p}\right) = 1.$$

26.2   Corollary. If $x^2 - my^2 = \pm p$ has a solution and $p \nmid m$, then

$$\left(\frac{m}{p}\right) = 1.$$

Unfortunately, the converse of Corollary 26.2 is false. Consider the Diophantine equation

$$x^2 + 5y^2 = 7.$$

This has no solution since $7 - 5y^2$ is not a square when $y = 0$ or 1, the only two values of y for which it is positive. Nevertheless

$$\left(\frac{-5}{7}\right) = \left(\frac{9}{7}\right) = 1.$$

We can prove a partial converse of Theorem 26.1 from which the converse of Corollary 4.2 follows for some special values of m. We start by trying to generalize the argument in Theorem 14.5.

26.3   Lemma. If $\left(\dfrac{m}{p}\right) = 1$, then there are integers x, y, and k such that $|k| \leq |m|$ and

$$x^2 - my^2 = kp. \tag{28}$$

Proof.   Since $\left(\dfrac{m}{p}\right) = 1$, there is an integer z such that

$$z^2 \equiv m \ (p).$$

Thue's theorem (14.4) implies the existence of integers x and y, not both 0, satisfying $|x| < \sqrt{p}$, $|y| < \sqrt{p}$, and

$$zy \equiv x \ (p).$$

Then

$$my^2 \equiv x^2 \ (p)$$

so

$$x^2 - my^2 = kp$$

for some integer k.   Finally,

$$
\begin{aligned}
|k|p = |kp| &= |x^2 - my^2| \\
&\leq x^2 + |m|y^2 \\
&< p + |m|p \\
&= p(1 + |m|).
\end{aligned}
$$

Since both k and m are integers, this implies $|k| \leq |m|$.

Now our problem is reduced to deciding when k can be taken to be ± 1 in Eq. (28).   When $m = -5$ and $p = 7$, the best we can manage is $k = 2$:

$$3^2 + 5 \cdot 1^2 = 2 \cdot 7.$$

Sometimes we are luckier.

26.4   Theorem. $x^2 + 2y^2 = p$ has a solution for the odd prime p if and only if $p \equiv 1$ or 3 modulo 8, or, equivalently, $\left(\dfrac{-2}{p}\right) = 1$.

Proof.   "Only if" follows immediately from Corollary 26.2 and Theorem 22.5.

To prove the converse suppose $\left(\dfrac{-2}{p}\right) = 1$, and apply Lemma 26.3; there are integers x, y, and k with $0 \le |k| \le 2$, and

$$x^2 + 2y^2 = kp.$$

We are done if $k = 1$. Since k is clearly positive, only the case $k = 2$ remains. Suppose

$$x^2 + 2y^2 = 2p.$$

Then x is even; suppose $x = 2u$. Then

$$4u^2 + 2y^2 = 2p$$

implies

$$y^2 + 2u^2 = p$$

so p is the sum of a square and twice a square, as desired.

26.5 Theorem. Let p be a prime other than 2 or 3. Then

$$x^2 + 3y^2 = p$$

has a solution if and only if $p \equiv 1(6)$, or, equivalently, $\left(\dfrac{-3}{p}\right) = 1$.

Proof. As before, " only if " follows from Corollary 26.2 and Theorem 25.4. Conversely, if $\left(\dfrac{-3}{p}\right) = 1$, then Lemma 26.3 implies the existence of integers x, y, and k for which

$$x^2 + 3y^2 = kp$$

where $0 \le k \le 3$. As before, $k = 0$ is impossible.

If $k = 1$ the theorem is true for p. Suppose $k = 3$. Then $3 \mid x$; say $x = 3u$. Then

$$9u^2 + 3y^2 = 3p$$

so

$$y^2 + 3u^2 = p$$

and the theorem is true for p.

We shall be done when we show $k = 2$ cannot occur. Suppose

$$x^2 + 3y^2 = 2p. \tag{29}$$

Since $p \equiv 1(6)$, $p \equiv 1(3)$, so reducing Eq. (29) modulo 3 yields

$$x^2 \equiv 2p \equiv 2 \ (3).$$

But this is impossible because 2 is not a residue of 3.

The next natural problem is Eq. (24) when $m = -4$ and p is an odd prime. We have already studied that problem, because

$$x^2 + 4y^2 = x^2 + (2y)^2.$$

Thus if

$$x^2 + 4y^2 = p \tag{30}$$

has a solution $\langle x, y \rangle$, then $\langle x, 2y \rangle$ solves

$$u^2 + v^2 = p. \tag{31}$$

Conversely, suppose p is an odd prime and $\langle u, v \rangle$ solves Eq. (31). Then just one of u and v is even; say $v = 2y$. Then $\langle u, y \rangle$ solves Eq. (30). Thus (30) and (31) are essentially equivalent problems. In general we need consider Eq. (25) only for square free integers m.

We have already remarked on the intractability of $m = -5$. The cases $m = -6$ and $m = -7$ are considered in Problems 27.7, 27.8, and 27.9.

So far we have discussed Eq. (25) only when $m < 0$. Then the existence of a solution for any particular prime p can be decided by trial and error in finitely many steps, for $|y|$ must be less than $\sqrt{p/|m|}$. This argument clearly fails for $m > 0$. Nevertheless we can make some progress for a few positive values of m.

26.6 Theorem. Let p be an odd prime. Then

$$x^2 - 2y^2 = p \tag{32}$$

has a solution if and only if $p \equiv \pm 1(8)$, or, equivalently, $\left(\dfrac{2}{p}\right) = 1$.

Proof. The start of the proof is familiar. If Eq. (32) has a solution, then $\left(\dfrac{2}{p}\right) = 1$ (Corollary 26.2), and hence $p \equiv \pm 1(8)$ (Theorem 24.3).

Conversely, suppose $p \equiv \pm 1(8)$. Then $\left(\dfrac{2}{p}\right) = 1$, so we may apply Lemma 26.3. There are integers x and y such that

$$x^2 - 2y^2 = kp$$

for $k = 0, \pm 1$, or ± 2; $k = 0$ is impossible because $\sqrt{2}$ is irrational. If $k = 1$, we are done. Suppose $k = \pm 2$. Then $x = 2u$ is even, so

$$4u^2 - 2u^2 = \pm 2p.$$

Therefore

$$y^2 - 2u^2 = \mp p.$$

Thus if $k = -2$ we are done, while the case $k = 2$ reduces to $k = -1$, the only case we have yet to consider. Suppose

$$x^2 - 2y^2 = -p.$$

We shall see in Section 30 why the following trick works. Let

$$u = x + 2y$$
$$v = x + y.$$

Then

$$
\begin{aligned}
u^2 - 2v^2 &= (x + 2y)^2 - 2(x + y)^2 \\
&= x^2 + 4xy + 4y^2 - 2x^2 - 4xy - 2y^2 \\
&= -x^2 + 2y^2 \\
&= p.
\end{aligned}
$$

This same trick also shows that $x^2 - 2y^2 = n$ has a solution if and only if $x^2 - 2y^2 = -n$ has one.

27. PROBLEMS

27.1 Let $p > 3$ be prime. Investigate the sum and product modulo p of the residues a of p between 0 and p. This problem and Problem 15.9 are related, though independent.

27.2 Can a primitive root for a prime $p > 2$ be a residue of p?

27.3 Show that the polynomial $(x^2 + 1)(x^4 - 4)$ has a root modulo p for every prime p but no integral root.

27.4 Prove that p is a Fermat prime (Problem 6.17) if and only if every non-residue of p is a primitive root for p. Prove that 3 is a primitive root for every Fermat prime > 3.

27.5 Compute $\left(\dfrac{503}{773}\right)$ and $\left(\dfrac{501}{773}\right)$. (The integers 501, 503, and 773 are prime.)

27.6 Prove: $\left(\dfrac{2}{p}\right) = (-1)^{(p^2 - 1)/8}$.

27.7 Prove: $\left(\dfrac{-7}{p}\right) = 1$ if and only if $p \equiv 1, 2,$ or $4(7)$.

27.8 Prove that

$$x^2 + 7y^2 = p$$

has a solution for the odd prime p if and only if $p \equiv 1, 2,$ or 4 modulo 7. *Hint*: Mimic the proof of Theorem 26.5. Only $k = 2$ or 4 should require ingenuity. Then try computing modulo 4 or 8.

27.9* Investigate the converse of Corollary 26.2 for $m = -6$.

27.10* Use Theorem 25.3 to decide when

$$x^2 - 3y^2 = \pm p$$

has a solution. Watch the signs.

27.11* Prove Theorems 25.4 and 25.3 by applying the lemma of Gauss directly, as in the proof of Theorem 24.3.

27.12 Show that 10 is a nonresidue of a prime p (different from 2 and 5) if and only if $p \equiv \pm 7, \pm 11, \pm 17,$ or $\pm 19(40)$. Deduce that these primes are nines numbers. (See Problem 22.3.)

27.13* Find a prime simultaneously of each of the forms $x^2 + y^2, x^2 + 2y^2, \ldots,$ $x^2 + 10y^2$. (See Elementary Problem E 1922, *Amer. Math. Monthly*, **73**, (1966), 891. The solution is in the same *Monthly*, **75**, (1968), 193.)

27.14 Suppose $p^\alpha \mid n!$; $p^{\alpha+1} \nmid n!$. Show $\alpha = [n/p] + [n/p^2] + [n/p^3] + \cdots$.

6

Quadratic Number Fields

In Section 14 and Problem 15.11 we solved the Diophantine equation $x^2 - y^2 = n$. The heart of that solution was the fact that $x^2 - y^2 = (x - y)(x + y)$. The analogous identity $(x - y\sqrt{m})(x^2 + y\sqrt{m}) = x^2 - my^2$ motivates this chapter, in which we study a class of rings each of which contains $\mathbf{Z}$ as a subring. Our results about arithmetic in these rings will generalize some number theory in $\mathbf{Z}$ and shed much light on the Diophantine equation

$$x^2 - my^2 = n \tag{1}$$

for some values of m.

28. THE FIELD $Q(\sqrt{m})$

Let m be an integer other than 0 or 1 which has no square factors; we say then that m is *square free*. Only such integers concern us in this chapter. We write $\sqrt{m}$ for the positive square root when $m > 0$ and for $i\sqrt{-m}$ when $m < 0$. Problems 6.12 and 15.6 each imply $\sqrt{|m|}$ is irrational. The symbol $\mathbf{Q}$ denotes the field of rational numbers, $\mathbf{C}$ the field of complex numbers. Let

$$\mathbf{Q}(\sqrt{m}) = \{a + b\sqrt{m} \mid a, b \in \mathbf{Q}\}$$
$$= \mathbf{Q} + \mathbf{Q}\sqrt{m}.$$

We have lost nothing by assuming m square free, for if $m' = n^2 m$, then $\mathbf{Q}(\sqrt{m'}) = \mathbf{Q}(\sqrt{m})$.

28.1   Theorem.   The set $\mathbf{Q}(\sqrt{m})$ is a subfield of $\mathbf{C}$.

Proof. $\mathbf{Q}(\sqrt{m})$ is obviously closed under subtraction.   Since

$$(a + b\sqrt{m})(c + d\sqrt{m}) = (ac + mbd) + (ad + bc)\sqrt{m} \tag{2}$$

it is closed under multiplication as well.   Thus $\mathbf{Q}(\sqrt{m})$ is a subring of $\mathbf{C}$. To show it is a field suppose

$$0 \neq \alpha = a + b\sqrt{m} \in \mathbf{Q}(\sqrt{m}).$$

We know α has an inverse in $\mathbf{C}$; we must prove $\alpha^{-1} \in \mathbf{Q}(\sqrt{m})$.   To that end we "rationalize the denominator," an elementary trick we shall formalize in Theorem 28.5 and systematically exploit throughout this chapter:

$$\frac{1}{a + b\sqrt{m}} = \frac{1}{a + b\sqrt{m}} \cdot \frac{a - b\sqrt{m}}{a - b\sqrt{m}}$$

$$= \frac{a}{a^2 - mb^2} - \frac{b}{a^2 - mb^2}\sqrt{m}. \tag{3}$$

The denominator $a^2 - mb^2$ cannot be zero since m is square free.

When $m > 0$, $\mathbf{Q}(\sqrt{m})$ is called a *real* quadratic field, when $m < 0$, a *complex* one.   The complex fields turn out to be much simpler than the real ones.

28.2   Lemma.   The representation $\alpha = a + b\sqrt{m}$ of an element of $\mathbf{Q}(\sqrt{m})$ is unique.   Thus α is rational if and only if $b = 0$.

Proof.   Suppose $\alpha = a + b\sqrt{m} = c + d\sqrt{m}$.   If $b \neq d$, then

$$\sqrt{m} = \frac{a - c}{d - b}$$

which would contradict the irrationality of $\sqrt{m}$.   Thus $b = d$ and so, of course, $a = c$.

28.3 Definition. The *conjugate* $\bar{\alpha}$ and the *norm* $N(\alpha)$ of

$$\alpha = a + b\sqrt{m} \in \mathbf{Q}(\sqrt{m})$$

are defined by

$$\bar{\alpha} = a - b\sqrt{m}$$
$$N(a) = \alpha\bar{\alpha} = a^2 - mb^2.$$

This definition makes sense only because α determines a and b (Lemma 28.2). When $m < 0$ the conjugate of α is its traditional complex conjugate, and $N(\alpha)$ is the square of the usual norm of α regarded as a complex number. The norm links the study of the field $\mathbf{Q}(\sqrt{m})$ to the Diophantine equation $x^2 - my^2 = n$, for that equation can be solved if and only if n is the norm of an element $x + y\sqrt{m}$ of $\mathbf{Q}(\sqrt{m})$ for which x and y are integers.

28.4 Lemma. The conjugation map $\alpha \rightsquigarrow \bar{\alpha}$ is a field automorphism of $\mathbf{Q}(\sqrt{m})$. That is

$$\overline{\alpha + \beta} = \bar{\alpha} + \bar{\beta} \tag{4}$$

and

$$\overline{\alpha\beta} = \bar{\alpha}\bar{\beta}. \tag{5}$$

Moreover

$$\bar{\bar{\alpha}} = \alpha \tag{6}$$

and $\alpha = \bar{\alpha}$ if and only if α is rational.

Proof. Equation (4) is obvious. Equation (5) is true because Eq. (2) remains true when $\sqrt{m}$ is replaced by $-\sqrt{m}$. Equation (6) is also obvious; moreover it implies that the conjugation map is its own inverse and hence is one to one and onto. Finally, $\alpha = \bar{\alpha}$ if and only if $b = 0$, so the last assertion follows from Lemma 28.2.

28.5 Theorem.

(a) $N(\alpha\beta) = N(\alpha)N(\beta)$.
(b) $N(\alpha)$ is rational.
(c) $N(\alpha) = 0$ if and only if $\alpha = 0$.

(d) $N(\alpha) = \alpha^2$ if and only if α is rational.

(e)   If $\alpha \neq 0$, then

$$\frac{1}{\alpha} = \frac{\bar{\alpha}}{N(\alpha)}.$$

Proof.

(a) $N(\alpha\beta) = \alpha\beta\overline{\alpha\beta} = \alpha\beta\bar{\alpha}\bar{\beta} = \alpha\bar{\alpha}\beta\bar{\beta} = N(\alpha)N(\beta)$.

(b)   Part (b) is obvious.

(c)   Clearly $N(0) = 0$. If $N(\alpha) = \alpha\bar{\alpha} = 0$, then $\alpha = 0$ or $\bar{\alpha} = 0$, which means $\alpha = 0$.

(d) $N(\alpha) = \alpha\bar{\alpha} = \alpha^2$ if and only if $\alpha = \bar{\alpha}$, that is, if and only if α is rational.

(e)   Part (e) is obvious; cross multiply.   Note that (c) implies $N(\alpha) \neq 0$.

Part (e) of Theorem 28.5 is the promised formalization of the trick we used to prove Theorem 28.1.   Part (a) is the most useful part.   It shows that the norm is multiplicative, or, equivalently,

$$N\colon \mathbf{Q}(\sqrt{m})^* \to \mathbf{Q}^*$$

is a group homomorphism.

29. ALGEBRAIC INTEGERS

The *characteristic polynomial P* of $\alpha = a + b\sqrt{m} \in \mathbf{Q}(\sqrt{m})$ is defined by

$$\begin{aligned} P(x) &= (x - \alpha)(x - \bar{\alpha}) \\ &= x^2 - (\alpha + \bar{\alpha})x + \alpha\bar{\alpha}. \end{aligned} \tag{7}$$

The polynomial P has rational coefficients since $1, \alpha + \bar{\alpha} = 2a$ and $N(\alpha)$ are rational.

29.1 Definition.   The number $\alpha \in \mathbf{Q}(\sqrt{m})$ is an *algebraic integer* if and only if its characteristic polynomial has integer, not just rational, coefficients. Let $\mathbf{A}(m)$ be the set of algebraic integers in $\mathbf{Q}(\sqrt{m})$.   Thus $\alpha \in \mathbf{A}(m)$ if and only if $\alpha + \bar{\alpha}$ and $N(\alpha)$ are ordinary integers.   Elements of $\mathbf{A}(-1)$ are called *Gaussian integers*.

We shall call the elements of $\mathbf{Z}$ *rational* integers whenever the adjective is required to avoid confusion.   Every rational integer n is an algebraic integer since whatever value m may have, $n \in \mathbf{Z}$ will have characteristic polynomial

$x^2 - 2nx + n^2$. No other rational numbers are algebraic integers. That follows from Problem 15.5. The next lemma provides another proof.

29.2 Lemma. $A(m) \cap Q = Z$; $\overline{A(m)} = A(m)$.

Proof. We have just noted that $Z \subset A(m)$. Of course $Z \subset Q$. To prove the reverse inclusion suppose α is rational. Then $N(\alpha) = \alpha^2$ is a rational integer only if α is one.

Since α and $\bar{\alpha}$ have the same characteristic polynomial, one is in $A(m)$ when the other is. Thus $\overline{A(m)} = A(m)$.

There are, of course, algebraic integers which are not rational. The number $\sqrt{m}$ is in $A(m)$ since its characteristic polynomial is $x^2 - m$.

The number $1 + 2i$ is in $A(-1)$ since $1 + 2i + \overline{(1 + 2i)} = 2$, and $N(1 + 2i) = 1^2 + 2^2 = 5$. These examples suggest the next lemma.

29.3 Lemma. When a and b are (rational) integers, $\alpha = a + b\sqrt{m} \in A(m)$. That is, $Z + Z\sqrt{m} \subseteq A(m)$.

Proof. $\alpha + \bar{\alpha} = 2a \in Z$ and $N(\alpha) = a^2 - mb^2 \in Z$.

Lest the reader hastily jump to a false conclusion, note that

$$\omega = -\frac{1}{2} + \frac{\sqrt{-3}}{2} \tag{8}$$

is an integer in $A(-3)$ since

$$\bar{\omega} + \omega = -1 \tag{9}$$

and

$$N(\omega) = \omega\bar{\omega} = (\tfrac{1}{2})^2 - (-3)(\tfrac{1}{2})^2 = 1. \tag{10}$$

The number ω is extremely important. Let us digress briefly now to compute some of its properties for use later.

29.4 Lemma. The characteristic polynomial of ω is

$$P(x) = x^2 + x + 1 \tag{11}$$

so

$$\omega^2 + \omega + 1 = \bar{\omega}^2 + \bar{\omega} + 1 = 0. \tag{12}$$

The three cube roots of 1 in **C** are 1, ω, and ω^2.

Proof. Equations (9) and (10) imply that Eq. (11) gives the characteristic polynomial P of ω. Since

$$x^3 - 1 = (x - 1)P(x) = (x - 1)(x - \omega)(x - \bar{\omega})$$

we have found the cube roots of 1. Note that

$$\omega^{-1} = \omega^2 = \bar{\omega}. \tag{13}$$

The conclusion to which you might have jumped, the converse of Lemma 29.3, is true only two thirds of the time. Since m is a square free integer, $4 \nmid m$. Thus $m \equiv 1, 2,$ or 3 modulo 4.

29.5   Theorem.   If $m \not\equiv 1(4)$, then $\alpha = a + b\sqrt{m}$ is in $\mathbf{A}(m)$ if and only if a and b are rational integers. If $m \equiv 1(4)$, then $a + b\sqrt{m} \in \mathbf{A}(m)$ if and only if $2a$ and $2b$ are rational integers of the same parity.

Proof.   Let $\alpha = a + b\sqrt{m} \in \mathbf{A}(m)$. Then $2a \in \mathbf{Z}$ and

$$4N(\alpha) = 4(a^2 - mb^2) = (2a)^2 - m(2b)^2 \in \mathbf{Z}$$

so $m(2b)^2 \in \mathbf{Z}$. Since m has no square factors, it follows that $2b \in \mathbf{Z}$. Moreover $4 \mid (2a)^2 - m(2b)^2$. If $2b$ is even, then $4 \mid (2a)^2$, so $2a$ is even. If $2a$ is even, then $4 \mid m(2b)^2$, but $4 \nmid m$, so $2b$ is even too. Thus $2a \equiv 2b$ (2).

Next we show a and b must themselves be rational integers unless $m \equiv 1(4)$. This together with Lemma 29.3 will complete the first part of the theorem. Suppose $a \notin \mathbf{Z}$. Since $2a \in \mathbf{Z}$, $2a$ is odd. Then $2b$ is also odd, so $(2a)^2 \equiv (2b)^2 \equiv 1(4)$. But $(2a)^2 - m(2b)^2 \equiv 0(4)$, so $m \equiv 1(4)$. An analogous argument shows $b \notin \mathbf{Z} \Rightarrow m \equiv 1(4)$.

Finally, we must show that if $m \equiv 1(4)$ and $2a \equiv 2b(2)$ in **Z**, then $\alpha = a + b\sqrt{m} \in \mathbf{A}(m)$. Clearly $\alpha + \bar{\alpha} = 2a$ is an integer. Let us check the norm:

$$N(a + b\sqrt{m}) = a^2 - mb^2$$
$$= \tfrac{1}{4}((2a)^2 - m(2b)^2). \tag{14}$$

Since $2a \equiv 2b(2)$ and $m \equiv 1(4)$, the factor in parentheses on the right in Eq. (14) will always be divisible by 4.

Note that when $m \equiv 1(4)$, there are "more" algebraic integers than those given by Lemma 29.3. This is the anomaly we first discovered in $A(-3)$: $-3 \equiv 1(4)$, so $\omega = -\frac{1}{2} + \frac{1}{2}\sqrt{-3} \in A(-3)$. Here $2a = -1$ and $2b = 1$; $1 \equiv -1(2)$.

When $m \equiv 2$ or $3(4)$, $A(m) = Z + Z\sqrt{m}$; we shall always write algebraic integers as $a + b\sqrt{m}$ where a and b are rational integers. This form is awkward when $m \equiv 1(4)$; we should have to allow a and b to be halves of odd integers as well. Here is another alternative description of $A(m)$. Let $\xi = \xi_m$ be defined by

$$\xi = \sqrt{m} \quad \text{if} \quad m \not\equiv 1\,(4) \tag{15}$$

$$\xi = -\frac{1}{2} + \frac{\sqrt{m}}{2} \quad \text{if} \quad m \equiv 1\,(4). \tag{16}$$

The minus sign in Eq. (16) is not traditional. We include it so that $\xi_3 = \omega$. It does no harm for other values of m.

29.6 Lemma. $A(m) = \{a + b\xi \mid a, b \in Z\} = Z + Z\xi$.

Proof. When $m \not\equiv 1(4)$, this is simply Theorem 29.5. Suppose $m \equiv 1(4)$. Then for any a and b

$$\alpha = a + b\xi = a + b\left(-\frac{1}{2} + \frac{\sqrt{m}}{2}\right)$$

$$= a - \frac{b}{2} + \frac{b}{2}\sqrt{m}.$$

Theorem 29.5 says $\alpha \in A(m)$ if and only if

$$2\left(a - \frac{b}{2}\right) = 2a - b$$

and

$$2\left(\frac{b}{2}\right) = b$$

are integers of the same parity.   That happens if and only if a and b are integers.   In particular, $\xi \in A(m)$.

We shall need the following facts about ξ when $m \equiv 1(4)$.

$$\xi^2 = \left(-\frac{1}{2} + \frac{\sqrt{m}}{2}\right)^2$$

$$= \frac{1+m}{4} - \frac{\sqrt{m}}{2}$$

$$= \frac{m-1}{4} - \xi. \tag{17}$$

$$\bar{\xi} = -\frac{1}{2} - \frac{\sqrt{m}}{2}$$

$$= -1 - \xi.$$

$$N(\xi) = \xi\bar{\xi} \tag{18}$$

$$= \left(-\frac{1}{2}\right)^2 - m\left(\frac{1}{2}\right)^2$$

$$= \frac{1-m}{4} \tag{19}$$

which is an integer because $4 \mid m - 1$.   Observe that Eqs. (17), (18), and (19) generalize (9), (10), (12), and (13).   Finally,

$$N(a + b\xi) = (a + b\xi)\overline{(a + b\xi)}$$

$$= a^2 + ab(\xi + \bar{\xi}) + b^2\xi\bar{\xi}$$

$$= a^2 - ab + b^2\left(\frac{1-m}{4}\right) \tag{20}$$

which is an integer when a and b are rational integers.

29.7 Theorem. $A(m)$ is a subring of $Q(\sqrt{m})$.

Proof.   Represent the algebraic integers in the form established by Lemma 29.6.   Then it is clear that sums, differences, and rational integral multiples of algebraic integers are again algebraic integers.   Now note that $\xi^2 \in A(m)$

either because $\xi^2 = m$ (if $m \not\equiv 1(4)$) or by virtue of Eq. (17) (if $m \equiv 1(4)$). Thus $\mathbf{A}(m)$ is closed under multiplication because

$$(a + b\xi)(c + d\xi) = ac + (b + d)\xi + bd\xi^2.$$

This proof of Theorem 29.7 is short but unsatisfactory. It ought not be necessary first to identify $\mathbf{A}(m)$ explicitly, as we did in Theorem 29.5, in order to show that it is closed under subtraction and multiplication. The reader is invited to try to find a more straightforward proof. Problem 41.4 may help.

30. NORMS AND UNITS

Since

$$x^2 + y^2 = (x + iy)(x - iy)$$

a rational integer is a sum of two squares if and only if it is the norm of a Gaussian integer. The consequences and generalizations of that simple observation will occupy us for much of this chapter. When m is a square free integer let $\mathbf{B}(m)$ be the set of norms of nonzero algebraic integers of $\mathbf{Q}(\sqrt{m})$. Since norms of algebraic integers are integers, $\mathbf{B}(m) \subseteq \mathbf{Z}$. Since $1 \in \mathbf{A}(m)$, $1 = N(1) \in \mathbf{B}(m)$.

30.1 Lemma. If $m \not\equiv 1(4)$, then $n \in \mathbf{B}(m)$ if and only if the Diophantine equation

$$x^2 - my^2 = n \neq 0 \tag{21}$$

has a solution.

Proof. Equation (21) says just

$$N(x + y\sqrt{m}) = n.$$

The lemma then follows from Theorem 29.5, which says $x + y\sqrt{m} \in \mathbf{A}(m)$ if and only if x and y are integers.

Thus, for example, an odd prime is in $\mathbf{B}(-1)$ if and only if it is congruent to 1 modulo 4 (Theorem 14.5) and is in $\mathbf{B}(2)$ if and only if it is congruent to ± 1 modulo 8 (Theorem 26.6).

As usual the situation is more complicated when $m \equiv 1(4)$.

30.2   Lemma.   If $m \equiv 1(4)$, then the following three statements are equivalent.

(a) $n \in \mathbf{B}(m)$.

(b)   The Diophantine equation

$$x^2 - xy + \left(\frac{1-m}{4}\right)y^2 = n \neq 0 \tag{22}$$

has a solution.

(c)   The Diophantine equation

$$x^2 - my^2 = 4n \neq 0 \tag{23}$$

has a solution.

Proof.   The equivalence of (a) and (b) follows easily from Lemma 29.6 and Eq. (20).   Suppose (a) is true, so that

$$n = N(a + b\sqrt{m})$$

where $0 \neq a + b\sqrt{m} \in \mathbf{A}(m)$.   Then $2a$ and $2b$ are integers, so

$$4n = 4(a^2 - mb^2) = (2a)^2 - m(2b)^2$$

and hence Eq. (23) has a solution.   Conversely, suppose (c); let $\langle x, y \rangle$ solve Eq. (23).   Since $m \equiv 1(4)$, m is odd.   But $4n$ is even, so Eq. (23) implies x and y have the same parity.   Then $\alpha = (x/2) + (y/2)\sqrt{m} \in \mathbf{A}(m)$ (Theorem 29.5) and

$$N(\alpha) = \frac{x^2}{4} - m\frac{y^2}{4} = n$$

so $n \in \mathbf{B}(m)$.

Although it is no longer logically necessary, we can prove the equivalence of (b) and (c) directly by consulting the proof of Lemma 29.6.   If $\langle x, y \rangle$ solves Eq. (23), then $\langle 2x - y, y \rangle$ solves Eq. (22); while if $\langle u, v \rangle$ solves Eq. (22), then $(u + v)/2 \in \mathbf{Z}$ and $\langle (u + v)/2, v \rangle$ solves Eq. (23).

If $m < 0$, then $\mathbf{B}(m)$ contains only positive integers.   Since $N(x + y\sqrt{m})$ $= N(|x| + |y|\sqrt{m})$ is an increasing function of $|x|$ and $|y|$, all the elements of $\mathbf{B}(m)$ less than some fixed k may be found by computing $N(x + y\sqrt{m})$ for $|x| < \sqrt{k}$ and $|y| < \sqrt{k/|m|}$, where x and y are either integers or appropriate half integers, depending on m modulo 4. We used this fact implicitly in the example following Corollary 26.2 and in Section 1, where we found the first few integers which are sums of two squares. That is $\mathbf{B}(-1)$ begins:

$$1, 2, 4, 5, 8, 9, 10, 13, 16, 17, 18, 20, \dots . \tag{24}$$

An initial segment of $\mathbf{B}(-5)$ can be similarly computed:

$$1, 4, 5, 6, 9, 14, 20, 21, 25, 29, 30, \dots . \tag{25}$$

30.3 Lemma.   If b_1 and b_2 are in $\mathbf{B}(m)$ so is $b_1 b_2$.

Proof.   There are algebraic integers α_1 and α_2 such that $b_1 = N(\alpha_1)$ and $b_2 = N(\alpha_2)$. Then $\alpha_1 \alpha_2 \in \mathbf{A}(m)$, so

$$b_1 b_2 = N(\alpha_1 \alpha_2) \in \mathbf{B}\,(m).$$

We have used Theorem 29.7 and Part (a) of Theorem 28.5, the multiplicativity of the norm.

We can now see what motivated the trick used to finish the proof of Theorem 26.6. There we had to show that $-p \in \mathbf{B}(2)$ implied $p \in \mathbf{B}(2)$. That implication now follows from Lemma 30.3 and the fact that

$$N(1 + \sqrt{2}) = 1^2 - 2 \cdot 1^2 = -1 \tag{26}$$

so that $-1 \in \mathbf{B}(2)$.   What we did then was to set

$$u + v\sqrt{2} = (1 + \sqrt{2})(x + y\sqrt{2})$$

and verify that the norm is multiplicative in this special case.

In the special case $m = -1$ Lemma 30.3 says that a sum of two squares times a sum of two squares is again such a sum (Problem 6.3).   That follows directly from Diophantos' identity

$$(a^2 + b^2)(c^2 + d^2) = (ac - bd)^2 + (ad + bc)^2 \tag{27}$$

which, unmotivated, might be hard to guess.   What we have done and are doing is to see how such identities naturally occur in the proper abstract algebraic context.

Lemma 30.3 may be restated as: $\mathbf{B}(m)$ is a subsemigroup of $\mathbf{Z}^*$, the semi-group of nonzero integers under multiplication.   Most of the rest of this chapter concerns divisibility in $\mathbf{B}(m)$ and in $\mathbf{A}(m)^*$, the multiplicative semigroup of nonzero algebraic integers of $\mathbf{Q}(\sqrt{m})$. These semigroups are related by the norm, which is a semigroup homomorphism from $\mathbf{A}(m)^*$ onto $\mathbf{B}(m)$. The reader unfamiliar with the definitions of divisibility, units, associates, and primes in semigroups like $\mathbf{A}(m)^*$ and $\mathbf{B}(m)$ will find them in Appendix 1.

For example, $1 + i$ divides $2 + 3i$ in the Gaussian integers because $(1 + i) \times (2 + i) = 2 + 3i$. We then write $(1 + i) \,|\, (2 + 3i)$. We see from the sequence (24) that $2 \,|\, 18$ in $\mathbf{B}(-1)$. Note however that $6 \notin \mathbf{B}(-1)$, so of course it cannot divide 18 there.

30.4 Lemma. If $\alpha \,|\, \beta$ in $\mathbf{A}(m)^*$, then $N(\alpha) \,|\, N(\beta)$ in $\mathbf{B}(m)$.

Proof. If $\alpha \,|\, \beta$, then there is a $\gamma \in \mathbf{A}(m)$ such that $\alpha\gamma = \beta$. Then $N(\alpha)N(\gamma) = N(\beta)$, so $N(\alpha) \,|\, N(\beta)$.

The converse of Lemma 30.4 is false. Suppose $m = -77$. Then

$$N(3) = 9 \,|\, 81 = N(2 + \sqrt{-77}) \tag{28}$$

in $\mathbf{B}(-77)$, but $3 \nmid 2 + \sqrt{-77}$ in $\mathbf{A}(-77)$ because $(2 + \sqrt{-77})/3$ is not an algebraic integer.

30.5 Lemma. The algebraic integer $\alpha \,|\, N(\alpha)$ in $\mathbf{A}(m)$.

Proof. Here we regard $N(\alpha)$ as a rational integral element of $\mathbf{A}(m)$ rather than as an element of $\mathbf{B}(m)$. Since $\bar{\alpha} \in \mathbf{A}(m)$ and $\alpha\bar{\alpha} = N(\alpha)$, the lemma is proved.

Recall that $\mu \in \mathbf{A}(m)$ is a *unit* if and only if it divides 1, that is, if and only if $\mu\nu = 1$ for some $\nu \in \mathbf{A}(m)$. The units are the invertible elements of $\mathbf{A}(m)$; they form a group (Appendix 1).

The number i is a unit in $\mathbf{A}(-1)$ because $i(-i) = 1$. Equation (13) shows that ω is a unit in $\mathbf{A}(-3)$. The number $\mu = \frac{1}{2} + \frac{1}{2}\sqrt{5}$ is a unit in $\mathbf{A}(5)$ since

$$\left(\frac{1}{2} + \frac{\sqrt{5}}{2}\right)\left(-\frac{1}{2} + \frac{\sqrt{5}}{2}\right) = 1. \tag{29}$$

The factors on the left in Eq. (29) are algebraic integers because $5 \equiv 1(4)$.

30.6 Lemma. The algebraic integer μ is a unit if and only if $N(\mu) = \pm 1$, that is, $N(\mu)$ is a unit in $\mathbf{B}(m)$.

Proof. If $\mu v = 1$ in $\mathbf{A}(m)$, then $N(\mu)N(v) = 1$ in $\mathbf{B}(m)$. Thus $N(\mu)$ is invertible in $\mathbf{B}(m)$, hence in $\mathbf{Z}$, so $N(\mu) = \pm 1$.

Conversely, if

$$N(\mu) = \mu\bar{\mu} = \pm 1$$

then $\pm\bar{\mu}$ is the inverse of μ in $\mathbf{A}(m)$.

A unit μ of $\mathbf{A}(m)$ is *proper* if $N(\mu) = 1$, *improper* if $N(\mu) = -1$. Note that $N(-1) = 1$, so -1 is always a proper unit. There are no improper units when $m < 0$. Equation (26) shows $1 + \sqrt{2}$ is an improper unit in $\mathbf{A}(2)$.

30.7 Lemma. The ring $\mathbf{A}(3)$ has no improper unit.

Proof. If

$$N(x + y\sqrt{3}) = x^2 - 3y^2 = -1$$

then $x^2 \equiv -1(3)$, which is impossible. (We have just solved Problem 6.6.) Lemmas 30.1, 30.2, and 30.6 imply that finding the units in $\mathbf{A}(m)$ is equivalent to solving the Diophantine equations

$$x^2 - my^2 = \pm 1 \tag{30}$$

when $m \not\equiv 1(4)$ or to solving

$$x^2 - my^2 = \pm 4 \tag{31}$$

when $m \equiv 1(4)$. These are easy to handle when $m < 0$.

30.8 Theorem. The units in $\mathbf{A}(-1)$ are ± 1 and $\pm i$. The units in $\mathbf{A}(-3)$ are ± 1, $\pm\omega$, and $\pm\omega^2$. When $m < 0$ and $m \neq -1$, -3, the only units in $\mathbf{A}(m)$ are ± 1.

Proof. Suppose $m = -1$. Then

$$x^2 + y^2 = \pm 1$$

implies $|x| = 1$ and $y = 0$, or $x = 0$ and $|y| = 1$. Suppose $m = -3 \equiv 1(4)$. Then

$$x^2 + 3y^2 = 4$$

implies $|x| = 2$ and $y = 0$, or $|x| = |y| = 1$. The case $-x = y = 1$ corresponds to the unit ω and the solution $N(\omega) = 1$. The reader can check the other cases.

If $m < -1$, then $\langle \pm 1, 0 \rangle$ are the only solutions to Eq. (30). If $m < -1$ and $m \not\equiv 1(4)$, then $m < -3$ so $\langle \pm 1, 0 \rangle$ are the only solutions to Eq. (31).

Finding the units in $\mathbf{A}(m)$ when $m > 0$ is much harder. We discuss the problem in the next section.

31. UNITS IN REAL FIELDS: PELL'S EQUATION

We proved in Theorem 30.8 that $\mathbf{A}(m)$ has only finitely many units when $m < 0$. That is false when $m > 0$. We know $\mu = 1 + \sqrt{2}$ is a unit in $\mathbf{A}(2)$ (Eq. (26)). Since μ is real and greater than 1, its powers μ^n are different units, so the group of units of $\mathbf{A}(2)$ is infinite. In fact, every unit of $\mathbf{A}(2)$ is $\pm \mu^n$ for some n. That fact is typical of $\mathbf{A}(m)$ for positive m.

For the remainder of this section m is a square free positive integer. Since $\mathbf{Q}(\sqrt{m})$ is a subfield of the field of real numbers, we will be able to use arguments based on the order structure of $\mathbf{A}(m)$.

Recall that $\mathbf{A}(m) = \mathbf{Z} + \mathbf{Z}\xi$, where $\xi = \sqrt{m}$ if $m \not\equiv 1(4)$, and

$$\xi = \frac{-1 + \sqrt{m}}{2}$$

if $m \equiv 1(4)$ (Lemma 29.6). Since the second case first occurs when $m = 5$, ξ is always greater than 0.

In $\mathbf{A}(m)$

$$\overline{a + b\xi} = \begin{cases} a - b\xi & m \not\equiv 1\,(4) \\ (a - 1) - b\xi & m \equiv 1\,(4). \end{cases} \tag{32}$$

31.1 Lemma. Let $\mu = a + b\xi$ be a unit other than ± 1 in $\mathbf{A}(m)$. Then just one of the four units $\pm \mu$, $\pm \bar{\mu}$ lies in each of the intervals $(-\infty, -1)$, $(-1, 0)$, $(0, 1)$, $(1, \infty)$. Moreover, $\mu > 1$ if and only if $a > 0$ and $b > 0$.

Proof. The four units $\pm \mu$, $\pm \mu^{-1}$ distribute themselves one each among the intervals in question. Since

$$\mu^{-1} = \frac{\bar{\mu}}{N(\mu)} = \pm \bar{\mu}$$

that is true of the units $\pm \mu$, $\pm \bar{\mu}$, so the first assertion is true.

Clearly $c + d\xi > 1$ if c and d are positive integers. Since $\mu \neq \pm 1$, we know $a \neq \pm 1$ and $b \neq 0$. Then Eq. (32) implies just one of the units $\pm \mu$, $\pm \bar{\mu}$ has two positive coefficients; that must be the one in $(1, \infty)$. Both parts of the lemma are false for arbitrary integers. For example, $73 + 5\sqrt{2} > 73 - 5\sqrt{2} > 1$.

31.2 Corollary. The interval $(1, M]$ contains only finitely many units.

Proof. There are only finitely many positive rational integers a, b for which $a + b\xi \leq M$.

Most of the work of this section will be in the proof of the next theorem; we shall postpone its proof until we have examined its consequences.

31.3 Theorem. There are integers $y \neq 0$ and x such that

$$x^2 - my^2 = 1$$

or, equivalently, $\mathbf{A}(m)$ has a proper unit other than ± 1.

Proof. See 31.9.

31.4 Theorem. There is a unit $\mu_0 \neq \pm 1$ in $\mathbf{A}(m)$ such that every unit is of the form $\pm \mu_0{}^n$ for some n.

Proof. Theorem 31.3 implies that there is a unit $\mu = x + y\sqrt{m} \neq \pm 1$. Lemma 31.1 implies we may assume $\mu > 1$. Then there are finitely many units in the interval $(1, \mu]$ (Corollary 31.2); let μ_0 be the least one. The theorem will be proved once we show every unit $v > 1$ is a positive power of μ_0, for then by Lemma 32.1, $\pm \mu_0{}^n$ will exhaust the units.

Suppose $v > 1$ is a unit. There is an $n > 0$ such that

$$0 < \mu_0^{n-1} < v \leq \mu_0{}^n. \tag{33}$$

Multiply (33) through by μ_0^{-n+1}:

$$0 < 1 < v\mu_0^{-n+1} \leq \mu_0. \tag{34}$$

But $v\mu_0^{-n+1}$ is a unit, and μ_0 is the least unit greater than 1, so

$$v\mu_0^{-n+1} = \mu_0$$

and thus

$$v = \mu_0{}^n$$

as desired.

31.5 Corollary. The group of units of $\mathbf{A}(m)$ is isomorphic to $\mathbf{Z}_2 \times \mathbf{Z}$.

Theorem 31.4 is a special case of Dirichlet's *unit theorem*, which asserts that in any number field $\mathbf{K}$ (we have defined and studied only the *quadratic number fields*) the group U of units in the ring of algebraic integers is isomorphic to $G \times \mathbf{Z}^k$ where G is a finite group and k is an integer which measures the extent to which $\mathbf{K}$ is real. For quadratic fields $k = 0$ when $m < 0$ (Theorem 30.8); $k = 1$ when $m > 0$ (Theorem 31.5).

31.6 Definition. The μ_0 constructed in Theorem 31.4 is the *fundamental unit* of $\mathbf{A}(m)$.

The algebraic integer $1 + \sqrt{2}$ is the fundamental unit of $\mathbf{A}(2)$ since it is a unit (Eq. (26)) and is clearly the least algebraic integer > 1 which has positive coefficients; $2 + \sqrt{3}$ is the fundamental unit of $\mathbf{A}(3)$ since $N(2 + \sqrt{3}) = 4 - 3 = 1$ and $1 + \sqrt{3}$ is not a unit.

This kind of argument shows that the fundamental unit can always be found in finitely many steps once some unit is known. But we can always find a unit by trying $y = 1, 2, \ldots$ in $1 + my^2$ and waiting for the result to be a perfect square. This procedure is not recommended. If $m = 31$, success first occurs when $y = 273$. Then $x = 1520$.

The hardest part of the argument is yet to come. The technique just outlined may be impractical, but Theorem 31.3 shows it is possible. We must now prove that theorem by finding a nontrivial solution to the Diophantine equation

$$x^2 - my^2 = 1. \tag{35}$$

Fermat challenged John Wallis and Lord Brouncker to solve Eq. (35). Though Fermat never published a solution, he probably knew one, for his challenge singled out the cases $m = 109$, 149, and 433, which require large values of x and y. Wallis eventually published Brouncker's answer to Fermat's challenge; this work fell into Euler's hands. He mysteriously credited the solution to Pell; Eq. (35) is therefore always called *Pell's equation*. For a more detailed history consult E. E. Whitford's *The Pell Equation*, Dissertation, Columbia University, New York, 1912.

The solution we give below is due to Dirichlet; it uses his pigeonhole principle (Lemma 14.3) several times and so is not an efficient algorithm for computing. Lagrange used the theory of continued fractions, which we shall not discuss, to give an elegant and efficient algorithm for finding not only a solution to Pell's equation but the one which minimizes $\alpha = x + y\sqrt{m}$ for $x > 1$ and $y > 0$; α will be the fundamental unit in $A(m)$ when $m \not\equiv 1(4)$ and $A(m)$ has no improper units. Otherwise it will be a power of the fundamental unit (see Problem 41.9). Moreover, if improper units exist, the continued fraction algorithm will find one. That is, it is good for solving the Diophantine equation $x^2 - my^2 = -1$.

Appendix 4 contains a table which lists the fundamental units of the rings $A(m)$ and the fundamental solutions to Pell's equation.

The argument which follows depends on the existence of "good" rational approximations to the irrational number $\sqrt{m}$, for if $\langle x, y \rangle$ satisfies Eq. (35) and $y \neq 0$, then

$$ x - y\sqrt{m} = \frac{1}{x + y\sqrt{m}} $$

so

$$ \left| \frac{x}{y} - \sqrt{m} \right| \leq \frac{1}{y(x + y\sqrt{m})} < \frac{1}{y^2}. \tag{36} $$

We shall look for pairs $\langle x, y \rangle$ which satisfy (36) and try to reverse the implications to find solutions to Eq. (35).

31.7 Lemma (Dirichlet). Let $\alpha > 0$ be irrational and M a positive integer. Then there are integers $0 < y \leq M$ and $x \geq 0$ such that

$$ |x - y\alpha| < \frac{1}{M}. \tag{37} $$

Proof. Recall that $[t]$ is the greatest integer less than or equal to t (Section 25) so that

$$ 0 \leq t - [t] < 1. $$

Divide the interval $[0, 1)$ into the M subintervals

$$ \left[0, \frac{1}{M} \right), \left[\frac{1}{M}, \frac{2}{M} \right), \ldots, \left[\frac{M-1}{M}, 1 \right) $$

each of length $1/M$. Then the pigeonhole principle (Lemma 14.3) implies that two of the $M + 1$ numbers

$$\alpha - [\alpha], 2\alpha - [2\alpha], \ldots, M\alpha - [M\alpha], (M + 1)\alpha - [(M + 1)\alpha]$$

must lie in the same interval. That is, there are integers y' and y'' for which $0 < y' < y'' \leq M + 1$ and

$$|([y''\alpha] - [y'\alpha]) - (y'' - y')\alpha| < \frac{1}{M}.$$

Let $x = [y''\alpha] - [y'\alpha]$ and $y = y'' - y' < M$.
Then

$$|x - y\alpha| < \frac{1}{M}$$

and we are done.

31.8 Lemma. Let $\alpha > 0$ be irrational. Then there are infinitely many pairs $\langle x, y \rangle$ of integers for which $y \neq 0$ and

$$\left| \frac{x}{y} - \alpha \right| < \frac{1}{y^2}. \tag{38}$$

Proof. That there is one such pair is obvious: Let $y = 1$ and $x = [\alpha]$. Suppose we have found n pairs $\langle x_i, y_i \rangle$. Let δ be the minimum of the n distances

$$|x_i - \alpha y_i|.$$

Since α is irrational, $\delta > 0$. Let M be an integer $> 1/\delta$. Then apply Lemma 31.7 to α and M to find x and $y \neq 0$ satisfying (37); $\langle x, y \rangle$ is not one of the pairs $\langle x_i, y_i \rangle$ since $1/M < \delta$. But $0 < y \leq M$, so (37) implies

$$\left| \frac{x}{y} - \alpha \right| < \frac{1}{My} \leq \frac{1}{y^2}.$$

Thus $\langle x, y \rangle$ is an $(n + 1)$st pair satisfying (38).

31.9 Proof of Theorem 31.3. We start by finding a k for which $N(\alpha) = k$ for infinitely $\alpha \in \mathbf{Z} + \mathbf{Z}\sqrt{m} \subseteq A(m)$. Suppose

$$\left| \frac{x}{y} - \sqrt{m} \right| < \frac{1}{y^2}.$$

Then

$$|N(x + y\sqrt{m})| = |x - y\sqrt{m}|\,|x + y\sqrt{m}|$$

$$= y^2 \left|\frac{x}{y} - \sqrt{m}\right|\left|\frac{x}{y} + \sqrt{m}\right|$$

$$< \left|\frac{x}{y} + \sqrt{m}\right|$$

$$= \left|\frac{x}{y} - \sqrt{m} + 2\sqrt{m}\right|$$

$$\le \left|\frac{x}{y} - \sqrt{m}\right| + 2\sqrt{m}$$

$$< \frac{1}{y^2} + 2\sqrt{m}$$

$$< 1 + 2\sqrt{m}.$$

Lemma 31.8 then implies $N(x + y\sqrt{m})$ is an integer between $-1 - 2\sqrt{m}$ and $1 + 2\sqrt{m}$ for infinitely many pairs $\langle x, y\rangle$, $y \ne 0$, so there is a rational integer k with $|k| < 1 + 2\sqrt{m}$ such that

$$N(\alpha) = N(x + y\sqrt{m}) = k \tag{39}$$

for infinitely many $\alpha \in \mathbf{Z} + \mathbf{Z}\sqrt{m}$.

Suppose $\alpha \ne \pm\beta$, and both satisfy Eq. (39). Then $\alpha\beta^{-1} \ne \pm 1$, and

$$N(\alpha\beta^{-1}) = kk^{-1} = 1$$

so $\alpha\beta^{-1}$ will be a nontrivial unit if it is an algebraic integer. Now

$$\alpha\beta^{-1} = \frac{\alpha\bar{\beta}}{N(\beta)} = \frac{\alpha\bar{\beta}}{k}$$

so $\alpha\beta^{-1}$ will be an algebraic integer if $k \mid \alpha\bar{\beta}$ in $\mathbf{A}(m)$. We shall produce $\alpha \ne \pm\beta$ with this property.

There are only k^2 possible values modulo $|k|$ for the pairs $\langle x, y\rangle$, so among the infinitely many α which satisfy (39) there are two, $\alpha \ne \pm\beta$, such that $k \mid \alpha - \beta$.

Then

$$k \mid (\alpha - \beta)\bar{\beta} = \alpha\bar{\beta} - \beta\bar{\beta} = \alpha\bar{\beta} - k$$

so $k \mid \alpha\bar{\beta}$ and we are done.

32. PRIMES

Recall that α and β are *associates* in $\mathbf{A}(m)^*$ if and only if each divides the other or, equivalently, $\alpha = \mu\beta$ for some unit μ. (See Definition 4 and Theorem 5 in Appendix 1.) For example, in $\mathbf{A}(-3)$

$$\lambda = 1 - \omega = \frac{3}{2} - \frac{\sqrt{-3}}{2}$$

and $\sqrt{-3}$ are associates since

$$\sqrt{-3} = -\omega\lambda \tag{40}$$

and ω is a unit in $\mathbf{A}(-3)$.

In $\mathbf{A}(2)$, $3 + \sqrt{2}$ and $75 + 53\sqrt{2}$ are associates since the quotient

$$\frac{75 + 53\sqrt{2}}{3 + \sqrt{2}} = \frac{(75 + 53\sqrt{2})(3 - \sqrt{2})}{7}$$

$$= \frac{119 + 84\sqrt{2}}{7}$$

$$= 17 + 12\sqrt{2}$$

$$= (1 + \sqrt{2})^4 \tag{41}$$

is a unit.

32.1 Lemma. If $\alpha \mid \beta$ in $\mathbf{A}(m)$ and $|N(\alpha)| = |N(\beta)|$, then α and β are associates.

Proof. We can write $\beta = \alpha\gamma$. Then

$$|N(\beta)| = |N(\alpha)|\,|N(\gamma)|$$

so $|N(\gamma)| = 1$. Lemma 30.6 implies γ is a unit, hence α and β are associates.

Recall that a nonunit π in $\mathbf{A}(m)^*$ (or $\mathbf{B}(m)$) is prime if and only if its only divisors are its associates and the units. Thus π is prime if and only if $\pi = \alpha\beta$ always implies α or β is a unit.

Since conjugation is an automorphism of the ring $\mathbf{A}(m)$, π is prime in $\mathbf{A}(m)$ if and only if $\bar{\pi}$ is prime.

Inspection of the sequences (24) and (25) shows that

$$2, 5, 9, 13, 17$$

are the first few primes in $\mathbf{B}(-1)$, while

$$4, 5, 6, 9, 14, 21, 29, \ldots \tag{42}$$

are prime in $\mathbf{B}(-5)$.

The next easy theorem gives a quick test for finding primes in $\mathbf{A}(m)$.

32.2 Theorem. If $N(\pi)$ is prime in $\mathbf{B}(m)$, then π is prime in $\mathbf{A}(m)$.

Proof. Suppose $\alpha \mid \pi$. Then $N(\alpha) \mid N(\pi)$ (Lemma 30.4), so either $N(\alpha) = \pm 1$, in which case α is a unit (Lemma 30.6), or $N(\alpha) = \pm N(\pi)$, in which case α and π are associates (Lemma 32.1). Thus π is prime.

32.3 Corollary. If $N(\pi)$ is prime in $\mathbf{Z}$, then π is prime in $\mathbf{A}(m)$.

Proof. An element of $\mathbf{B}(m)$ which is prime in $\mathbf{Z}$ is *a fortiori* prime in $\mathbf{B}(m)$.

32.4 Corollary. There are infinitely many primes in $\mathbf{A}(m)$ and $\mathbf{B}(m)$.

Proof. Since elements of $\mathbf{A}(m)$ with norms prime in $\mathbf{B}(m)$ are prime (Theorem 32.2), it suffices to prove the corollary for $\mathbf{B}(m)$. Let p be a rational prime. Then $p^2 = N(p) \in \mathbf{B}(m)$; p^2 will be prime in $\mathbf{B}(m)$ unless $\pm p \in \mathbf{B}(m)$, in which case $\pm p$ is prime in $\mathbf{B}(m)$. Since there are infinitely many rational primes, we are done.

32.5 Examples. Let us illustrate how Theorem 32.2 and Corollary 32.3 are used. Since $N(1 + 2i) = 5$, $1 + 2i$ is prime in $\mathbf{A}(-1)$.

There are, however, primes in $\mathbf{A}(m)$ whose norms are not rational primes. The converse of Corollary 32.3 is false; we really need the full strength of Theorem 32.2. To see that, note that $1 + \sqrt{-5}$ is prime in $\mathbf{A}(-5)$ because its norm is 6, which is prime in $\mathbf{B}(-5)$ (see (42)). Of course 6 is not prime in $\mathbf{Z}$. Note too for later reference that 2 and 3 are prime in $\mathbf{A}(-5)$ because their norms, 4 and 9, are prime in $\mathbf{B}(-5)$.

The converse of Theorem 32.2 is false too. Just after we proved Lemma 30.4, we observed that $81 = N(2 + \sqrt{-77})$ was not prime in $\mathbf{B}(-77)$. However, it is easy to see that $2 + \sqrt{-77}$ is prime in $\mathbf{A}(-77)$, for suppose α is a proper divisor of $2 + \sqrt{-77}$. Then $N(\alpha)$ is a proper divisor of 81 (Lemma 30.4), so $N(\alpha) = 3$, 9, or 27. But the Diophantine equation

$$x^2 + 77^2 y^2 = n$$

has no solutions when $n = 3$ or 27 and only the solutions $\langle \pm 3, 0 \rangle$ when $n = 9$. Thus the only possible proper divisors of $2 + \sqrt{-77}$ are $\alpha = \pm 3$, and neither of these does divide it. Hence $2 + \sqrt{-77}$ is prime in $\mathbf{A}(-77)$, but its norm is composite in $\mathbf{B}(-77)$.

32.6 Theorem. Every nonunit in $\mathbf{A}(m)^*$ (or $\mathbf{B}(m)$) is a product of primes.

Proof. The proof below is for $\mathbf{A}(m)$; the method is induction on $|N(\alpha)|$ for $\alpha \in \mathbf{A}(m)$. To construct the proof for $\mathbf{B}(m)$, simply replace $|N(\alpha)|$ by $|\alpha|$ throughout. (Compare the proof of Corollary 12 in Appendix 1.)

It is clear that every nonunit α with $|N(\alpha)| \le 2$ is prime and hence is a product of primes. Suppose every α with $1 < |N(\alpha)| < n$ is a product of primes, and suppose $|N(\gamma)| = n$. If γ is prime, it is a product of primes, and we are done. If γ is not prime, then $\gamma = \alpha\beta$ where neither α nor β is a unit. Thus $|N(\alpha)| > 1$ and $|N(\beta)| > 1$, so $|N(\alpha)| < |N(\gamma)| = n$, and $|N(\beta)| < |N(\gamma)| < n$. Our inductive hypothesis implies α and β are products of primes, so γ is too.

The natural question to raise next is that of the uniqueness of the factorization established in Theorem 32.6. Consider $\mathbf{A}(-5)$. There

$$6 = 2 \cdot 3 = (1 + \sqrt{-5})(1 - \sqrt{-5}). \tag{43}$$

We showed in 32.5 that 2, 3, and $1 + \sqrt{-5}$ are prime. Since $1 - \sqrt{-5}$ $= \overline{1 + \sqrt{-5}}$, it too is prime. However, $1 + \sqrt{-5}$ is not an associate of 2 or of 3 since its norm is different from $N(2)$ and from $N(3)$. Therefore 6 has been factored into a product of primes in two essentially different ways. That is unfortunate.

The structure of the domain $\mathbf{A}(m)$ and corresponding information about the Diophantine equation

$$x^2 - my^2 = n$$

is beyond the scope of this book unless factorization in $\mathbf{A}(m)$ is unique. In the next section we shall exhibit some values of m for which $\mathbf{A}(m)$ has that property. Then we proceed to some of its consequences.

33. EUCLIDEAN NUMBER FIELDS

We have twice used the theorem that an integral domain in which "long division" is possible enjoys unique factorization. We shall show now that it applies to $\mathbf{A}(m)$ for some small values of m. To save page turning we repeat here some definitions from Appendix 1.

33.1 Definition. An integral domain R is a *Euclidean domain* if it has a *Euclidean norm*, that is, a function $E : R^* \to \mathbf{Z}^+$ which satisfies

(a) $E(ab) = E(a)E(b)$.
(b) Given $a \neq 0$ and b in R there are elements q and $r \in R$ such that $b = aq + r$ and either $r = 0$ or $E(r) < E(b)$.

33.2 Theorem. A Euclidean domain is a unique factorization domain.

Proof. This is Theorem 16 of Appendix 1.

Now consider $\mathbf{A}(m)$. The obvious candidate for a Euclidean norm is the absolute value of the norm. Since

$$|N(\alpha\beta)| = |N(\alpha)|\,|N(\beta)|$$

condition (a) of Definition 33.1 is satisfied.

33.3 Lemma. The function $|N|$ is a Euclidean norm for $\mathbf{A}(m)$ if given $\alpha \neq 0$ and $\beta \in \mathbf{A}(m)$ there is a $\tau \in \mathbf{A}(m)$ such that

$$\left| N\!\left(\tau - \frac{\beta}{\alpha} \right) \right| < 1.$$

Proof. We need only check condition (b) in Definition 33.1. Given $\alpha \neq 0$ and β, let $q = \tau$ and $r = \beta - \alpha\tau \in \mathbf{A}(m)$. Then $\beta = \alpha q + r$, and

$$|N(r)| = |N(\beta - \tau)| = \left| N\!\left(\alpha\!\left(\frac{\beta}{\alpha} - \tau \right) \right) \right|$$

$$= |N(\alpha)| \left| N\!\left(\frac{\beta}{\alpha} - \tau \right) \right|$$

$$< |N(\alpha)|$$

so either $r = 0$ or $0 < |N(r)| < |N(\alpha)|$ as required. Note that $(\beta/\alpha) - \tau$ is probably not an algebraic integer; we have taken advantage of the fact that N is defined and multiplicative on all of $\mathbf{Q}(\sqrt{m})$.

33.4 Theorem. The function $|N|$ is a Euclidean norm for $\mathbf{A}(m)$ when $m = -2, -1, 2,$ or 3.

Proof. Suppose $\alpha \neq 0$ and β in $\mathbf{A}(m)$. Lemma 33.3 suggests that we look for an algebraic integer τ near the quotient $\beta/\alpha = u + v\sqrt{m}$ in $\mathbf{Q}(\sqrt{m})$.

Here u and v are rational numbers, so there are rational integers s and t such that

$$|u - s| \le \tfrac{1}{2}, \qquad |v - t| \le \tfrac{1}{2}.$$

Let $\tau = s + t\sqrt{m} \in \mathbf{A}(m)$. Then

$$\frac{\beta}{\alpha} - \tau = (u - s) + (v - t)\sqrt{m}$$

so

$$\left| N\!\left(\frac{\beta}{\alpha} - \tau\right) \right| = |(u - s)^2 - m(v - t)^2|$$

$$\le (u - s)^2 + |m|\,(v - t)^2$$

$$\le \frac{1}{4} + \frac{|m|}{4}. \tag{44}$$

The last quantity in (44) is strictly less than 1 if $m = -2, -1$, or 2. When $m = 3$, we must work a little harder. Since $-3 < 0$, the first inequality in (44) will be strict unless $s = u$ or $t = v$. In these cases one of the terms is zero; the other is clearly less than 1. Thus $|N((\beta/\alpha) - \tau)| < 1$ in all cases.

Even that patched up argument fails when $m = -3$ because no cancellation occurs inside the absolute value bars. Nevertheless $|N|$ is a Euclidean norm for -3 because $\mathbf{A}(-3)$ has "more" integers than $\mathbf{A}(3)$. That is, there is more freedom in choosing s and t. If we let $x + y\sqrt{m}$ correspond to the point $\langle x, y \rangle$ in the Cartesian plane, then when $m \not\equiv 1(4)$, $\mathbf{A}(m)$ consists of the lattice points (Section 25). When $m \equiv 1(4)$, the centers of the small squares formed by lattice points are also integers; they correspond to the numbers $(s + t\sqrt{m})/2$ where s and t are both odd integers (Theorem 29.5).

33.5 Theorem. The function $|N|$ is a Euclidean norm for $\mathbf{A}(m)$ when $m = -11, -7, -3, 5$, or 13.

Proof. Note that each of these m is congruent to 1 modulo 4. Suppose $\alpha \ne 0$ and β are in $\mathbf{A}(m)$. We must look for an algebraic integer near $\beta/\alpha = u + v\sqrt{m}$ in $\mathbf{Q}(\sqrt{m})$. First find the nearest rational integer t to the rational number $2v$. Then

$$\left| \frac{t}{2} - v \right| \le \frac{1}{4}.$$

Let s be the nearest integer to $2u$ which has the same parity as t. Then $|s - 2u| \le 1$ so

$$\left| \frac{s}{2} - u \right| \le \frac{1}{2}.$$

Since $m \equiv 1(4)$, $\tau = (s/2) + (t/2)\sqrt{m} \in A(m)$ (Theorem 2.8). Then

$$\left| N\left(\frac{\beta}{\alpha} - \tau \right) \right| = \left| \left(u - \frac{s}{2} \right)^2 - m\left(v - \frac{t}{2} \right)^2 \right|$$

$$\le \begin{cases} \dfrac{1}{4} + \dfrac{|m|}{16} & (m < 0) \\[2ex] \max\left\{ \dfrac{1}{4}, \dfrac{m}{16} \right\} & (m > 0) \end{cases}$$

$$< 1$$

for each of the five relevant values of m.

We have just proved that $A(m)$ is a Euclidean domain and hence a unique factorization domain, for $m = -11, -7, -3, -2, -1, 2, 3, 5,$ and 13. It is known that $|N|$ is a Euclidean norm for $A(m)$ if and only if $m = -1$, $\pm 2, \pm 3, 5, 6, \pm 7, \pm 11, 13, 17, 19, 21, 29, 33, 37, 41, 57,$ or 73. No other complex quadratic number fields are Euclidean (Problem 41.50). There may be values of $m > 0$ different from those listed above for which $A(m)$ has a Euclidean norm different from $|N|$. We do not know, though their number is known to be finite.

$A(m)$ is a unique factorization domain, but not necessarily Euclidean, for just the square free $m < 100$ in Table 2.

TABLE 2

-1	2	21	46	71
-2	3	22	47	73
-3	5	23	53	77
-7	6	29	57	83
-11	7	33	59	86
-19	11	37	61	89
-43	13	38	62	93
-67	17	41	67	94
-163	19	43	69	97

Note that there are exactly nine negative m in the table. Gauss knew these nine numbers; it has only recently been proved that there are no others (Stark, Harold, *Proc. Nat. Acad. Sci. U.S.A.*, **57**, (1967), 216–221). It is not known whether $A(m)$ enjoys unique factorization for infinitely many positive m.

34. CONSEQUENCES OF UNIQUE FACTORIZATION

Throughout this section, m is a square free integer for which $A(m)$ is a unique factorization domain. We gave a partial list of such m at the end of the last section (Table 2).

34.1 Definitions and Discussion. Let p be a rational prime. Then $B(m)$ may or may not contain $\pm p$. If neither p nor $-p$ is in $B(m)$, then $p^2 = N(p) \in B(m)$ is prime in $B(m)$, so p is prime in $A(m)$ (Theorem 32.2). When that occurs we call p *inertial*. Suppose p is not inertial. Then $\pm p$ is a norm, so for some $\pi \in A(m)$,

$$N(\pi) = \pi\bar{\pi} = \pm p \in B(m). \tag{45}$$

Then π and $\bar{\pi}$ are prime in $A(m)$ (Corollary 32.3), and Eq. (45) gives the unique factorization of $\pm p$ into primes in $A(m)$. When π and $\bar{\pi}$ are associates in $A(m)$, we say p *ramifies*, otherwise p *splits*. Thus p ramifies when it is an associate of the square of a prime in $A(m)$.

34.2 Examples. The rational prime 5 is inertial in $A(3)$. The equation

$$x^2 - 3y^2 = \pm 4 \cdot 5$$

has no solution in integers because $(\tfrac{3}{5}) = -1$ (Theorem 26.1). The prime 2 ramifies in $A(3)$ since

$$-2 = (1 + \sqrt{3})(1 - \sqrt{3})$$

where the factors on the right are associates:

$$1 + \sqrt{3} = (1 - \sqrt{3})(2 + \sqrt{3})$$

and $2 + \sqrt{3}$ is a unit in $A(3)$ (Definition 31.6). The prime 11 splits in $A(3)$ because

$$-11 = N(1 + 2\sqrt{3}) = (1 + 2\sqrt{3})(1 - 2\sqrt{3}).$$

The factors on the right are not associates since

$$\frac{1 + 2\sqrt{3}}{1 - 2\sqrt{3}} = \frac{(1 + 2\sqrt{3})^2}{-11} = \frac{13 + 4\sqrt{3}}{-11}$$

is not an algebraic integer.

We must allow the possibility $-p$ in Definition 34.1. In the preceding example we found that $-11 \in \mathbf{B}(3)$. However, $11 \notin \mathbf{B}(3)$, for if for some $\alpha \in \mathbf{A}(3)$

$$N(\alpha) = \alpha\bar{\alpha} = 11 = a^2 - 3b^2,$$

then $11 \equiv a^2(3)$. But

$$\left(\frac{11}{3}\right) = \left(\frac{2}{3}\right) = -1.$$

A similar argument shows $13 \in \mathbf{B}(3)$ while $-13 \notin \mathbf{B}(3)$. These facts prompted the warning in Problem 27.10. This difficulty can not arise when $m < 0$, for then $-p$ is never in $\mathbf{B}(m)$, or when $\mathbf{A}(m)$ has an improper unit, for then both or neither of $\pm p$ are norms. Moreover, no comparable trouble occurs with $-p^2$.

34.3   Lemma.   If $-p^2 \in \mathbf{B}(m)$ then $-1 \in \mathbf{B}(m)$.

Proof. Either $-p^2$ is prime in $\mathbf{B}(m)$ or $-p^2 = p(-p) \in \mathbf{B}(m)$. In both cases there is a prime π in $\mathbf{A}(m)$ such that both $N(\pi)$ and $-N(\pi)$ are in $\mathbf{B}(m)$. Then $-N(\pi) = N(\sigma)$ for some σ, and

$$\pi\bar{\pi} = -\sigma\bar{\sigma}.$$

Then $\pi \mid \sigma$ (or $\pi \mid \bar{\sigma}$), so $\pi\mu = \sigma$ for some $\mu \in \mathbf{A}(m)$, and

$$N(\pi)\, N(\mu) = N(\sigma)$$

implies

$$N(\mu) = -1 \in \mathbf{B}\,(m).$$

34.4   Definition   We shall say $\varepsilon = \pm 1$ is an *appropriate sign* for $n > 0$ when $\varepsilon n \in \mathbf{B}(m)$. Thus $+1$ is always appropriate for p^2; if -1 is appropriate for some p^2, then $-1 \in \mathbf{B}(m)$ and ± 1 are both appropriate for all $n \in \mathbf{B}(m)$.

We can decide which rational primes ramify or split using the unique factorization of $\mathbf{A}(m)$.

34.5 Theorem. Let p be an odd prime and $\mathbf{A}(m)$ a unique factorization domain. Then

$$p \text{ ramifies} \Leftrightarrow p \mid m \Leftrightarrow \left(\frac{m}{p}\right) = 0$$

$$p \text{ splits} \Leftrightarrow \left(\frac{m}{p}\right) = 1 \tag{46}$$

$$p \text{ is inertial} \Leftrightarrow \left(\frac{m}{p}\right) = -1.$$

Proof. Since the three conditions in each column of (46) are exclusive and exhaustive, it suffices to prove each of the implications $\Leftarrow$. We start with the last.

Suppose $\left(\frac{m}{p}\right) = -1$. Then the Diophantine equations

$$x^2 - my^2 = \pm p, \qquad x^2 - my^2 = \pm 4p$$

have no solutions (Theorem 26.1 and Corollary 26.2). Consequently $\pm p \notin \mathbf{B}(m)$ (Lemma 30.1 or 30.2), and p is inertial.

Consider the two remaining possibilities. If $p \mid m$, then $p \mid p^2 - m$; if $\left(\frac{m}{p}\right) = 1$, then the congruence $x^2 \equiv m(p)$ has a solution, so in both of these cases

$$p \mid x^2 - m = (x + \sqrt{m})(x - \sqrt{m})$$

for some $x \in \mathbf{Z}$. If p were prime in $\mathbf{A}(m)$, then it would divide either $x + \sqrt{m}$ or $x - \sqrt{m}$, but since $p \neq 2$, neither $(x + \sqrt{m})/p$ or $(x - \sqrt{m})/p$ is an algebraic integer. Therefore p is not prime in $\mathbf{A}(m)$. We need only decide now when it splits, and when it ramifies.

We know $\pm p \in \mathbf{B}(m)$, so there are integers a and b such that

$$\pi = a + b\sqrt{m} \in \mathbf{A}(m)$$

and

$$\pi\bar{\pi} = N(\pi) = \pm p = a^2 - mb^2$$

if $m \not\equiv 1(4)$, or

$$\pi = \frac{a + b\sqrt{m}}{2} \in \mathbf{A}\,(m)$$

and

$$4\pi\bar{\pi} = 4N(\pi) = \pm 4p = a^2 - mb^2$$

if $m \equiv 1(4)$. Then for either kind of m,

$$p \mid a \Leftrightarrow p \mid m. \qquad (47)$$

We wish to decide when p ramifies, or, equivalently, when $\pi/\bar{\pi}$ is a unit in $\mathbf{A}(m)$. But

$$\frac{\pi}{\bar{\pi}} = \frac{\pi^2}{N(\pi)} = \frac{\pi^2}{\pm p}$$

so p ramifies if and only if $p \mid \pi^2$.

We shall show

$$p \mid \pi^2 \Leftrightarrow p \mid a$$

which, coupled with (47), will prove the theorem. Now

$$(a^2 + mb^2) + 2ab\sqrt{m} = \begin{cases} \pi^2 & m \not\equiv 1\ (4) \\ 4\pi^2 & m \equiv 1\ (4) \end{cases}$$

so clearly

$$p \mid a \quad \text{implies} \quad p \mid \pi^2.$$

Finally, suppose $p \mid \pi^2$. Then $p \mid 2ab$, so p, which is odd, divides a or b. If $p \mid a$, we are done. If $p \mid b$, then p must also divide a since $p \mid a^2 + mb^2$, so we are done.

Note that we have achieved a long sought goal; we have proved the converse of Corollary 26.2 for some values of m. The case-by-case arguments using Thue's theorem for $m = -1$ (Theorem 14.5), $m = -2, -3, 2$ (Theorems 26.4, 26.5, and 26.6), and $m = -7, 3$ (Problems 27.8 and 27.10) are obsolete. Moreover, we now know the source of the counterexample following Corollary 26.2. The domain $\mathbf{A}(-5)$ does not enjoy unique factorization.

In Theorem 34.5 we considered only odd primes p; 2 must be treated separately.

34.6 Theorem. Suppose $A(m)$ a unique factorization domain. Then 2 ramifies in $A(m)$ if $m \not\equiv 1(4)$, splits if $m \equiv 1(8)$, and is inertial if $m \equiv 5(8)$.

Since we will not need this theorem later, its proof is relegated to Problem 41.17. Let us check it in a few cases:

$$2 = (1 + i)(1 - i) = -i(1 + i)^2$$

so 2 ramifies in the Gaussian integers $A(-1)$, as it should, because $-1 \equiv 3(4)$. Since $2 = (\sqrt{2})^2$, 2 ramifies in $A(2)$. In $A(-7)$

$$2 = \left(\frac{1 + \sqrt{-7}}{2}\right)\left(\frac{1 - \sqrt{-7}}{2}\right). \tag{48}$$

The factors in Eq. (48) are not associates since ± 1 are the only units (Theorem 30.8). Thus 2 splits in $A(-7)$, which verifies the assertion of Theorem 34.6 since $-7 \equiv 1(8)$.

Finally, $-3 \equiv 5(8)$, and 2 is inertial in $A(-3)$ because the Diophantine equation

$$x^2 + 3y^2 = 8$$

has no solution.

We saw in 32.5 that $1 + \sqrt{-5}$ is prime in $A(-5)$ because its norm is 6, a prime in $B(-5)$. The integer 6 is neither a rational prime nor the square of one. That anomaly does not occur when $A(m)$ enjoys unique factorization.

34.7 Theorem. If $A(m)$ is a unique factorization domain, and π is prime in $A(m)$, then $N(\pi) = \pm p$ or $\pm p^2$ for some rational prime p.

Proof. Since π divides the positive integer $|N(\pi)|$, there is a least positive integer p such that $\pi \mid p$ in $A(m)$. Suppose $p = rs$ in Z. Then $\pi \mid r$ or $\pi \mid s$ in $A(m)$. The minimality of p then tells us $p = r$ or $p = s$. That is p is prime in Z. Now

$$\pi \bar{\pi} = N(\pi) \mid N(p) = p^2$$

so

$$N(\pi) = \pm p \quad \text{or} \quad \pm p^2.$$

34.8 Corollary. The primes in $\mathbf{A}(m)$ are the inertial rational primes and the prime factors of the rational primes which ramify or split.

Proof. Suppose π prime in $\mathbf{A}(m)$. If $N(\pi) = \pm p$, then π is one of the prime factors of p. If $N(\pi) = \pm p^2$, then $\pi \mid \pm p^2$ and hence $\pi \mid p$ in $\mathbf{A}(m)$. But $|N(\pi)| = |N(p)|$ so π and p are associates (Lemma 32.1), and hence p is an inertial prime.

34.9 Theorem. When $\mathbf{A}(m)$ is a unique factorization domain, $\mathbf{B}(m)$ enjoys unique factorization. The primes in $\mathbf{B}(m)$ are the squares of the inertial rational primes and the rational primes which split or ramify, taken with appropriate signs.

Proof. If $n \in \mathbf{B}(m)$ is prime in $\mathbf{B}(m)$, then it is the norm of a prime π of $\mathbf{A}(m)$ (Theorem 32.2), so Corollary 34.8 implies $n = \pm p$ or $\pm p^2$ for a rational prime p. To show $\mathbf{B}(m)$ enjoys unique factorization we must show $n \mid a$ or $n \mid b$ whenever $n \mid ab$ in $\mathbf{B}(m)\mu$. (Appendix 1, Theorem 7.) Now in $\mathbf{A}(m)$

$$\pi \mid N(\pi) \mid n \mid ab = \alpha\bar{\alpha}\beta\bar{\beta}.$$

Since π is prime in $\mathbf{A}(m)$, it divides one of the four factors on the right; we may assume $\pi \mid \alpha$. Then $N(\pi) \mid N(\alpha)$. That is, $n \mid a$.

34.10 Corollary. If $\mathbf{A}(m)$ is a unique factorization domain, then $n \in \mathbf{B}(m)$ if and only if

$$n = \varepsilon p_1^{\alpha_1} \cdots p_k^{\alpha_k} q_1^{2\beta_1} \cdots q_r^{2\beta_r}$$

where the q_i are inertial primes, the p_i split or ramify, and ε is an appropriate sign. (See Problem 41.19.)

Recall that $\mathbf{B}(m)$ was defined simply as the set of norms of the integers $\mathbf{A}(m)$. Thus Corollary 34.10 together with Theorem 34.5 and Lemma 30.1 or 30.2 shows that when $\mathbf{A}(m)$ is a unique factorization domain we have discovered when the Diophantine equation

$$x^2 - my^2 = n$$

or the equations

$$x^2 - my^2 = 4n$$

and

$$x^2 - xy + \left(\frac{1-m}{4}\right)y^2 = n$$

can be solved.   We have done much more.   In the next few sections we shall show by example in some special cases how we can use Corollary 34.10 constructively to compute and count solutions to those equations.

35.   THE DIOPHANTINE EQUATION $x^2 + 2y^2 = n$

Let us collect and exploit what we know about $\mathbf{A}(-2)$ to solve the Diophantine equation

$$x^2 + 2y^2 = n. \tag{49}$$

The domain $\mathbf{A}(-2)$ is a Euclidean (Theorem 33.4) and hence a unique factorization domain.   The only units in $\mathbf{A}(-2)$ are ± 1 (Theorem 30.8). The prime 2 ramifies since $-2 = N(\sqrt{-2})$, and $\sqrt{-2}$ and $-\sqrt{-2}$ are associates.   An odd prime p is inertial if it is congruent to 5 or 7 modulo 8 and splits if congruent to 1 or 3 modulo 8 (Theorems 34.5 and 24.5 or Theorem 26.4).   Lemma 30.1 links these facts about $\mathbf{A}(-2)$ to the equation we are studying: $\langle x, y \rangle$ solves Eq. (49) if and only if

$$\tau = x + y\sqrt{-2} \in \mathbf{A}(-2)$$

and $N(\tau) = n$.   Moreover, if

$$\tau' = x' + y'\sqrt{-2}$$

then $\tau = \tau'$ if and only if $\langle x, y \rangle = \langle x', y' \rangle$ (Lemma 28.2), so there are as many solutions to Eq. (49) as there are integers in $\mathbf{A}(-2)$ with norm n.

We wish to count in a more sophisticated way.   We do not want to consider $\langle 1, 3 \rangle$ and $\langle 1, -3 \rangle$ different solutions to

$$x^2 + 2y^2 = 19.$$

35.1 Definition.   The solutions $\langle x, y \rangle$ and $\langle x', y' \rangle$ are *equivalent* solutions to Eq. (49) if and only if $x = \pm x'$ and $y = \pm y'$.

Fortunately, $\mathbf{A}(-2)$ has a built-in device which identifies equivalent solutions.

35.2 Lemma. The solution $\langle x, y \rangle$ is equivalent to $\langle x', y' \rangle$ if and only if $\tau = \pm\tau$ or $\pm\bar{\tau}$.

The proof is short and straightforward.

Now we can count the number of inequivalent solutions to Eq. (49) and simultaneously give an algorithm for finding them when they are known for n prime. The reader who solved Problem 15.11 discovered the technique the following proof exploits.

35.3 Theorem. The Diophantine equation

$$x^2 + 2y^2 = n \tag{50}$$

has a solution if and only if

$$n = 2^\alpha p_1^{\alpha_1} \cdots p_k^{\alpha_k} q_1^{2\beta_1} \cdots q_r^{2\beta_r} \tag{51}$$

where the p_i are rational primes congruent to 1 or 3 modulo 8, and the q_i are rational primes congruent to 5 or 7 modulo 8. When n is of this form Eq. (50) has

$$\left[\frac{(\alpha_1 + 1) \cdots (\alpha_k + 1) + 1}{2} \right]$$

inequivalent solutions.

Proof. The existence of solutions for just those n satisfying Eq. (51) is Corollary 34.10. Next we count solutions when they exist. We must find all $\tau \in \mathbf{A}(-2)$ of norm n. Begin by factoring n in $\mathbf{A}(-2)$. The p_i are the primes which split; let $p_i = \pi_i \bar{\pi}_i$ (see Definition 34.1). Then π_i and $\bar{\pi}_i$ are primes which are not associates. Since $2 = -(\sqrt{-2})^2$,

$$n = (-1)^\alpha (\sqrt{-2})^{2\alpha} \pi_1^{\alpha_1} \bar{\pi}_1^{\alpha_1} \cdots \pi_k^{\alpha_k} \bar{\pi}_k^{\alpha_k} q_1^{2\beta_1} \cdots q_r^{2\beta_r}$$

is the "unique" factorization of n into distinct primes in $\mathbf{A}(-2)$. Now suppose $N(\tau) = \tau\bar{\tau} = n$. What can τ be? If σ is prime in $\mathbf{A}(-2)$ and $\sigma^\gamma \mid \tau$, then $\bar{\sigma}^\gamma \mid \bar{\tau}$, so $\tau\bar{\tau}$ will equal n if and only if the prime factorization of τ is:

$$\tau = \pm(\sqrt{-2})^\alpha \pi_1^{\gamma_1} \bar{\pi}_1^{\alpha_1 - \gamma_1} \cdots \pi_k^{\alpha_k} \bar{\pi}_k^{\alpha_k - \gamma_k} q_1^{\beta_1} \cdots q_r^{\beta_r} \tag{52}$$

for some integers $\gamma_1, \ldots, \gamma_k$ with $0 \le \gamma_i \le \alpha_i$. The primes q_i and $\sqrt{-2}$ must appear to the same power in τ and $\bar{\tau}$ because $\bar{q}_i = q_i$ and $\sqrt{-2}$, while not quite $\sqrt{-2}$, is $-\sqrt{-2}$, its associate.

Since τ and $-\tau$ lead to equivalent solutions of Eq. (50), we must ignore the sign in Eq. (52) when we count. Different choices for the γ_i then yield

$$A = (\alpha_1 + 1) \cdots (\alpha_k + 1)$$

possible choices for τ, no two of which are associates. However, we also must avoid counting both τ and $\bar{\tau}$ since they too lead to equivalent solutions. Since $\pm \bar{\tau}$ results when we interchange γ_i and $\alpha_i - \gamma_i$ in the right member of Eq. (52), the solutions corresponding to $\gamma_1, \ldots, \gamma_k$ and $\alpha_1 - \gamma_1, \ldots, \alpha_k - \gamma_k$ are equivalent. Thus each solution to Eq. (50) has been counted twice except in the special case in which $\tau = \pm \bar{\tau}$, which can happen only when every α_i is even and $\gamma_i = \alpha_i/2$. Then that solution has been counted just once. Both cases are covered by the expression

$$\left[\frac{A + 1}{2} \right]$$

for the number of inequivalent solutions. If at least one α_i is odd, then A is even and

$$\left[\frac{A + 1}{2} \right] = \frac{A}{2}.$$

If all the α_i are even, that is, if n is a power of 2 times a square, then A is odd, and

$$\left[\frac{A + 1}{2} \right] = \frac{A - 1}{2} + 1,$$

the correct answer. There is an anomaly in the count when $n = 2^\alpha t^2$, for then $\langle 2^{[\alpha/2]}t, 0 \rangle$ or $\langle 0, 2^{[\alpha/2]}t \rangle$ solves Eq. (50) (depending on whether α is even or odd), and that solution is equivalent to one rather than to three others.

35.4 Example. When $n = 3$, there should be $[((1 + 1) + 1)/2] = [3/2] = 1$ solution, and in fact

$$1^2 + 2 \cdot 1^2 = 3$$

is obviously the only way to write 3 as the sum of a square and twice a square. In general when p is a rational prime congruent to 1 or 3 modulo 8, that is, when p splits, Theorem 35.3 implies

$$x^2 + 2y^2 = p$$

has exactly one solution in positive integers x and y. This uniqueness assertion is implicit in Definition 34.1 but is an addition to Theorem 26.4.

When $n = 18 = 2 \cdot 3^2$, the anomalous case occurs.

$$\left[\frac{(\alpha_1 + 1) + 1}{2}\right] = \left[\frac{(2 + 1) + 1}{2}\right] = \left[\frac{4}{2}\right] = 2.$$

The two solutions are

$$18 = 0^2 + 2 \cdot 3^2$$
$$= 4^2 + 2 \cdot 1^2.$$

35.5 Example. Suppose $n = 2475 = 3^2 \cdot 11 \cdot 5^2$. There are

$$\left[\frac{(2 + 1)(1 + 1) + 1}{2}\right] = \left[\frac{7}{2}\right] = 3$$

solutions, and Theorem 35.3 shows how to find them. We must first factor 3 and 11 in $\mathbf{A}(-2)$:

$$3 = (1 + \sqrt{-2})(1 - \sqrt{-2})$$
$$11 = (3 + \sqrt{-2})(3 - \sqrt{-2}).$$

Then

$$2475 = (1 + \sqrt{-2})^2(1 - \sqrt{-2})^2(3 + \sqrt{-2})(3 - \sqrt{-2}) \cdot 5^2.$$

Here

$$\alpha = 0, \quad \alpha_1 = 2, \quad \text{and} \quad \alpha_2 = \beta_1 = 1.$$

The inequivalent solutions to $N(\tau) = 2475$ correspond to $\gamma_1 = 0$, 1, and 2, $\gamma_2 = 0$. These yield

$$\tau = (1 - \sqrt{-2})^2(3 - \sqrt{-2}) \cdot 5$$
$$= (-7 - 5\sqrt{-2}) \cdot 5$$
$$= -35 - 25\sqrt{-2},$$
$$\tau = (1 + \sqrt{-2})(1 - \sqrt{-2})(3 - \sqrt{-2}) \cdot 5$$
$$= (3 - \sqrt{-2}) \cdot 15$$
$$= 45 - 15\sqrt{-2},$$

and

$$\tau = (1 + \sqrt{-2})^2(3 - \sqrt{-2}) \cdot 5$$
$$= (1 + 7\sqrt{-2}) \cdot 5$$
$$= 5 + 35\sqrt{-2}.$$

You can then check that, in fact,

$$2475 = (35)^2 + 2(25)^2 = 5^2(7^2 + 2 \cdot 5^2)$$
$$= (45)^2 + 2(15)^2 = 15^2(3^2 + 2 \cdot 1^2)$$
$$= 5^2 + 2(35)^2 \quad = 5^2(1^2 + 2 \cdot 7^2).$$

These are the only solutions to

$$x^2 + 2y^2 = 2475.$$

A close reading of this example hints at the possibility and desirability of counting solutions $\langle x, y \rangle$ for which $(x, y) = 1$, or is as small as possible. You are invited to try in Problem 41.33.

36. THE SUM OF TWO SQUARES

We are ready to answer all the questions raised in Section 1. We shall apply the methods of the last section and our knowledge of the Gaussian integers $\mathbf{A}(-1)$ to discover in how many ways an integer n may be written as a sum of two squares.

The domain $\mathbf{A}(-1)$ is Euclidean (Theorem 33.4) and hence a unique factorization domain; the units in $\mathbf{A}(-1)$ are ± 1 and $\pm i$ (Theorem 30.8). The prime 2 ramifies in $\mathbf{A}(-1)$ (34.6). The odd primes congruent to 1 modulo 4 split, the others are inertial. That we knew as early as Theorem 14.5. It is, of course, also a consequence of Theorem 34.5 and Theorem 13.2. The pair $\langle x, y \rangle$ solves the Diophantine equation

$$x^2 + y^2 = n \tag{53}$$

if and only if

$$\tau = x + iy \in \mathbf{A}(-1)$$

and $N(\tau) = n$. Moreover if

$$\tau' = x' + iy'$$

then $\tau = \tau'$ if and only if $\langle x, y \rangle = \langle x', y' \rangle$, so Eq. (53) has as many solutions as there are Gaussian integers of norm n. However that is too refined a way to count. The pairs $\langle 2, 1 \rangle$ and $\langle -1, 2 \rangle$ should not be counted as different solutions to

$$x^2 + y^2 = 5.$$

The difference between this section and the last is that Eq. (53) is symmetric in x and y, while Eq. (49) is not. That is why we considered $\mathbf{A}(-2)$ before $\mathbf{A}(-1)$.

36.1 Definition. Two solutions $\langle x, y \rangle$ and $\langle x', y' \rangle$ to Eq. (53) are *equivalent* if and only if $x = \pm x'$ and $y = \pm y'$ or $x = \pm y'$ and $y = \pm x'$. Thus there can be eight solutions equivalent to a given one. Fortunately, $\mathbf{A}(-1)$ has four units rather than two.

36.2 Lemma. The solution $\langle x, y \rangle$ is equivalent to $\langle x', y' \rangle$ if and only if $\tau' = \mu\tau$ or $\mu\bar{\tau}$ where μ is one of the four units ± 1, $\pm i$.

Again the proof is easy; so we proceed immediately to the analogue of Theorem 35.3.

36.3 Theorem. The Diophantine equation

$$x^2 + y^2 = n \tag{54}$$

has a solution if and only if

$$n = 2^\alpha p_1^{\alpha_1} \cdots p_k^{\alpha_k} q_1^{2\beta_1} \cdots q_r^{2\beta_r} \tag{55}$$

where the p_i are rational primes congruent to 1 modulo 4, and the q_i are rational primes congruent to 3 modulo 4. When n is of this form, Eq. (54) has

$$\left[\frac{(\alpha_1 + 1) \cdots (\alpha_k + 1) + 1}{2} \right]$$

inequivalent solutions.

Proof. We can almost copy the proof of Theorem 35.3. Corollary 34.10 implies that solutions exist for just those n satisfying Eq. (55). To count the solutions we look for Gaussian integers τ of norm n.

When $p_i \equiv 1(4)$ is a rational prime, it splits. Let $p_i = \pi_i \bar{\pi}_i$; π_i and $\bar{\pi}_i$ are different primes. Since

$$2 = (1 + i)(1 - i)$$

$$= -i(1 + i)^2,$$

$$n = (-i)^\alpha (1 + i)^{2\alpha} \pi_1^{\alpha_1} \bar{\pi}_1^{\alpha_1} \cdots \pi_k^{\alpha_k} \bar{\pi}_k^{\alpha_k} q_1^{2\beta_1} \cdots q_r^{2\beta_r}$$

in $\mathbf{A}(-1)$, and

$$\tau = \mu(1 + i)^\alpha \pi_1^{\gamma_1} \bar{\pi}_1^{\alpha_1 - \gamma_1} \cdots \pi_k^{\gamma_k} \bar{\pi}_k^{\alpha_k - \gamma_k} q_1^{\beta_1} \cdots q_r^{\beta_r}$$

for some unit μ and rational integers $\gamma_1, \ldots, \gamma_k$ with $0 \le \gamma_i \le \alpha_i$. We ignore μ since changing it does not change the equivalence class of the solution τ produces. The

$$A = (\alpha_1 + 1) \cdots (\alpha_k + 1)$$

different choices for the γ_i yield different Gaussian integers no two of which are associates. The search for conjugate pairs among them proceeds exactly as in the proof of Theorem 35.3, and gives the same result. Strictly speaking we seek not just conjugate pairs but pairs τ, τ' for which $\bar{\tau}$ and τ' are associates. That means only that we must write μ where $\pm$ appears in the proof of Theorem 35.3.

To illustrate Theorem 36.3 let us verify the statement made in Section 1 that 325 is the least integer representable three ways as a sum of two squares. If n is so representable, we must have

$$3 \le \frac{(\alpha_1 + 1) \cdots (\alpha_k + 1) + 1}{2} < 4$$

so

$$5 \le (\alpha_1 + 1) \cdots (\alpha_k + 1) < 7.$$

This occurs only when $k = 1$, $\alpha_1 = 4$, 5, or $k = 2$, $\alpha_1 = 2$, $\alpha_2 = 1$. The two smallest primes congruent to 1 modulo 4 are 5 and 13, so the smallest integers for which these sequences of exponents occur are 5^4 and $5^2 \cdot 13$. Of these the second, 325, is smaller.

Other questions about sums of squares are raised in Problems 41.34–41.38.

37. A(−3) *AND RELATED DIOPHANTINE EQUATIONS*

Our next task is to study the Diophantine equations

$$x^2 + 3y^2 = n \tag{56}$$

$$x^2 + 3y^2 = 4n \tag{57}$$

and

$$x^2 - xy + y^2 = n. \tag{58}$$

The domain $\mathbf{A}(-3)$ is Euclidean and hence a unique factorization domain (Theorem 33.5).   The units in $\mathbf{A}(-3)$ are ± 1, $\pm\omega$, and $\pm\omega^2$, where

$$\omega = \frac{-1 + \sqrt{-3}}{2} \tag{59}$$

(Theorem 30.8).   The prime 2 is inertial (34.6); 3 ramifies since

$$-3 = (\sqrt{-3})^2.$$

Every other rational prime p is congruent to 1 or 5 modulo 6; the former split, the latter are inertial (Theorems 25.4 and 34.5 or Theorem 26.5 and Lemma 30.2).   We shall use this information first to study Eq. (58).   The pair $\langle x, y \rangle$ solves Eq. (58) if and only if

$$\tau = x + y\omega \in \mathbf{A}(-3)$$

and

$$N(\tau) = n$$

(Lemma 29.6, Eq. (20), and Lemma 30.2).   Moreover, if

$$\tau' = x' + y'\omega$$

then $\tau = \tau'$ if and only if $\langle x, y \rangle = \langle x', y' \rangle$, so Eq. (58) has as many solutions as there are integers of $\mathbf{A}(-3)$ of norm n.

As before, however, we wish to consider some solutions equivalent to others.   Clearly $\langle x, y \rangle$, $\langle -x, -y \rangle$, and $\langle y, x \rangle$ should not all be counted. Note however, that $\langle x, y \rangle$ and $\langle x, -y \rangle$ are not both solutions.   To define equivalence we shall reverse the roles of Lemma 36.2 and Definition 36.1.

37.1 Definition. Algebraic integers τ and τ' yield equivalent solutions $\langle x, y \rangle$ and $\langle x', y' \rangle$ of Eq. (58) if and only if $\tau' = \mu\tau$ or $\mu\bar{\tau}$ where $\mu = \pm 1$, $\pm\omega$, $\pm\omega^2$.

37.2 Lemma. The solutions equivalent to $\langle x, y \rangle$ are

$$\langle x, y \rangle, \quad \langle -y, x - y \rangle, \quad \langle y - x, -x \rangle,$$
$$\langle x - y, -y \rangle, \quad \langle y, x \rangle, \quad \langle -x, y - x \rangle$$

and their negatives.

Proof. These correspond in order to τ, $\omega\tau$, $\omega^2\tau$, $\bar{\tau}$, $\omega\bar{\tau}$, $\omega^2\bar{\tau}$ and their negatives. The verification of that assertion, which we omit, is an easy exercise using Eqs. (9), (10), (12), and (13).

Note that Definition 37.1 was almost forced upon us. We wished to have $\langle x, y \rangle$ and $\langle y, x \rangle$ equivalent. These correspond to τ and $\omega\bar{\tau}$; it would be irrational then not to call τ and $\omega\tau$ equivalent, which leads to the unsuspected equivalent solution

$$\langle -y, x - y \rangle.$$

We have now arranged the definition of equivalent solutions so that the analogue of Theorems 35.3 and 36.3 is true.

37.3 Theorem. The Diophantine equation

$$x^2 - xy + y^2 = n \tag{60}$$

has a solution if and only if

$$n = 2^{2\beta}3^\alpha p_1^{\alpha_1} \cdots p_k^{\alpha_k}q_1^{2\beta_1} \cdots q_r^{2\beta_r} \tag{61}$$

where the p_i are rational primes congruent to 1 modulo 6, and the q_i are rational primes congruent to 5 modulo 6. When n is of this form, there are

$$\left[\frac{(\alpha_1 + 1) \cdots (\alpha_k + 1) + 1}{2} \right]$$

unequivalent solutions.

The proof is just like those of Theorems 35.3 and 36.3.

We proved in Lemma 30.2 that the Diophantine equation

$$u^2 + 3v^2 = 4n \tag{62}$$

has solutions if and only if Eq. (60) has. In fact, we noted there that if $\langle x, y \rangle$ solves Eq. (60), then

$$\langle u, v \rangle = \langle 2x - y, y \rangle \tag{63}$$

solves Eq. (62). We can use this fact together with Lemma 37.2 to prove the next theorem.

37.4 Theorem. The Diophantine equation

$$w^2 + 3z^2 = n \tag{64}$$

has a solution if and only if $n \in \mathbf{B}(-3)$, or, equivalently, n satisfies Eq. (61).

Proof. If the integers w and z solve Eq. 64, then $w + z\sqrt{-3} \in \mathbf{A}(-3)$, so

$$N(w + z\sqrt{-3}) = n \in \mathbf{B}(-3).$$

Conversely, suppose $n \in \mathbf{B}(-3)$. Then we can find a solution $\langle x, y \rangle$ to Eq. (60). At least one of the equivalent solutions $\langle x, y \rangle$, $\langle y, x \rangle$, or $\langle -y, x - y \rangle$ has an even second component, so we may assume y even from the start. Then both u and v in Eq. (63) are even, and $\langle u, v \rangle$ solves Eq. (62). We can then divide Eq. (62) through by 4 to produce a solution to Eq. (64).

We leave as an exercise the problem of counting the solutions to Eqs. (62) and (64) (Problem 41.39).

38. THE DIOPHANTINE EQUATION $x^2 - 2y^2 = n$

It should be clear by now that information about

$$x^2 - 2y^2 = n \tag{65}$$

can be deduced from knowledge of the arithmetic of the unique factorization domain $\mathbf{A}(2)$ (Theorem 33.4). The prime 2 ramifies in $\mathbf{A}(2)$ since it is the square of the algebraic integer $\sqrt{2}$. An odd prime p splits if it is congruent to ± 1 modulo 8, otherwise it is inertial (Theorem 26.6 or Theorems 24.3 and 34.5). Since $2 > 0$, $\mathbf{A}(2)$ has infinitely many units; they are given by

$$\pm(1 + \sqrt{2})^k = \pm\mu_0{}^k$$

for $k \in \mathbf{Z}$ (Theorem 31.4 and Definition 31.6). The pair $\langle x, y \rangle$ solves Eq. (65) if and only if

$$\tau = x + y\sqrt{2} \in A(2)$$

and $N(\tau) = n$. Moreover, if

$$\tau' = x' + y'\sqrt{2}$$

then $\tau = \tau'$ if and only if $\langle x, y \rangle = \langle x', y' \rangle$. Thus Eq. (65) has as many solutions as there are algebraic integers of $A(2)$ with norm n. That statement is not much use in view of the next lemma.

38.1 Lemma. For all integral k, $N(\tau) = N(\pm(1 + \sqrt{2})^{2k}\tau)$, so that if Eq. (65), has one solution it has infinitely many.

Proof. The norm is multiplicative and $N(\pm(1 + \sqrt{2})^{2k}) = 1$.

For example, $\langle 3, 1 \rangle$ solves

$$x^2 - 2y^2 = 7.$$

Therefore so does $\langle x, y \rangle$ when

$$x + y\sqrt{2} = \pm(3 + \sqrt{2})(1 + \sqrt{2})^{2k}.$$

The first few solutions in this sequence are $\langle 13, 9 \rangle$, $\langle 5, -3 \rangle$, and $\langle 75, 53 \rangle$, corresponding to $k = 1, -1$, and 2. (See Eq. (41).)

We cannot hope to count solutions, but we can count equivalence classes.

38.2 Definition. The solutions $\langle x, y \rangle$ and $\langle x', y' \rangle$ to Eq. (65) are *equivalent* if and only if $\tau = \pm\mu_0^{2k}\tau$ or $\pm\mu_0^{2k}\bar{\tau}$ for some k. We omit the routine argument which shows that this does in fact define an equivalence relation on the set of solutions.

38.3 Theorem. The Diophantine equation

$$x^2 - 2y^2 = n \tag{66}$$

has a solution if and only if

$$n = \pm 2^\alpha p_1^{\alpha_1} \cdots p_k^{\alpha_k} q_1^{2\beta_1} \cdots q_r^{2\beta_r} \tag{67}$$

where the p_i are rational primes congruent to ± 1 modulo 8 and the q_i are rational primes congruent to ± 3 modulo 8. When n is of this form, there are

$$\left[\frac{(\alpha_1 + 1) \cdots (\alpha_k + 1) + 1}{2}\right]$$

infinite equivalence classes of solutions.

Proof. Corollary 34.10 shows that solutions exist if and only if n satisfies Eq. (67); both signs are appropriate (Definition 34.4) since $-1 = N(1 + \sqrt{2}) \in \mathbf{B}(2)$. Let $p_i = \pi_i \bar{\pi}_i$ in $\mathbf{A}(2)$. Then

$$n = \pm(\sqrt{2})^{2\alpha} \pi_1^{\alpha_1} \bar{\pi}_1^{\alpha_1} \cdots \pi_k^{\alpha_k} \bar{\pi}_k^{\alpha_k} q_1^{2\beta_1} \cdots q_r^{2\beta_r}$$

in $\mathbf{A}(2)$. Every solution $\langle x, y \rangle$ of Eq. (66) corresponds to a divisor τ of n in $\mathbf{A}(2)$. Then

$$\tau = \pm \mu_0^{\gamma} (\sqrt{2})^{\alpha} \pi_1^{\gamma_1} \bar{\pi}_1^{\alpha_1 - \gamma_1} \cdots \pi_k^{\gamma_k} \bar{\pi}_k^{\alpha_k - \gamma_k} q_1^{\beta_1} \cdots q_r^{\beta_r} \tag{68}$$

for some integers $\gamma, \gamma_1, \ldots, \gamma_k$ with $0 \le \gamma_i \le \alpha_i$. Since

$$n = \tau \bar{\tau} = (\mu_0 \bar{\mu}_0)^{\gamma} n = (-1)^{\gamma} n$$

it follows that γ is even. Then the solution which corresponds to $\mp \mu_0^{-\gamma} \tau$ is equivalent to $\langle x, y \rangle$, so we may ignore the sign and the first factor in Eq. (68). There are thus

$$A = (\alpha_1 + 1) \cdots (\alpha_k + 1)$$

different choices for τ no two of which are associates. The usual argument shows that the conjugate of each such τ occurs once among the A choices, so that there are $A/2$ equivalence classes unless τ and $\bar{\tau}$ are associates for some τ. That happens just once when A is odd; in that case there are $((A - 1)/2) + 1 = (A + 1)/2$ equivalence classes of solutions. It is clear from Definition 38.2 that each equivalence class is infinite.

39. *QUADRATIC FORMS AND QUADRATIC NUMBER FIELDS*

We have succeeded in studying successfully the Diophantine equations

$$x^2 - y^2 = n$$
$$x^2 + 2y^2 = n$$
$$x^2 + y^2 = n$$
$$x^2 + 3y^2 = n$$
$$x^2 - xy + y^2 = n$$
$$x^2 - 2y^2 = n$$

because in each case the appropriate ring $A(m)$ of integers is a unique factorization domain.   The first of these equations is particularly simple because $A(1) = Z$.   We solved it in Section 14 and Problems 15.10 and 15.11.   In general, the study of

$$ax^2 + bxy + cy^2 = n \tag{69}$$

is intimately related to the arithmetic of $A(m)$ where

$$m = \begin{cases} b^2 - 4ac & \text{if } b \text{ is odd} \\ \left(\dfrac{b}{2}\right)^2 - ac & \text{if } b \text{ is even.} \end{cases}$$

In order to study Eq. (69) we need a theory which works even when unique factorization fails in $A(m)$.

The theory of ideals developed by Kummer and Dedekind in order to study the Fermat conjecture (Chapter 7) does that job.   We shall briefly sketch its outlines for the ring $A(-5)$.   The details do not belong in a book titled "Elementary . . . ."

We know that in $A(-5)$

$$6 = 2 \cdot 3 = (1 + \sqrt{-5})(1 - \sqrt{-5}) \tag{70}$$

where 2, 3, $1 + \sqrt{-5}$, and $1 - \sqrt{-5}$ are all prime (Example 32.5).   Suppose we did not believe that these were primes because we felt factorization should be unique.   We might then imagine that 2, 3, $1 + \sqrt{-5}$, and $1 - \sqrt{-5}$ had "ideal" prime factors, much as -1 has an "imaginary" square root.

Suppose π were one of the ideal prime divisors of 6.   Then we should have

$$\pi \,|\, 2 \quad \text{or} \quad \pi \,|\, 3$$
$$\text{and} \quad \pi \,|\, 1 + \sqrt{-5} \quad \text{or} \quad \pi \,|\, 1 - \sqrt{-5}.$$

For definiteness suppose the first alternative in each case.   Then π would be a common divisor of 2 and $1 + \sqrt{-5}$.   But we know how to look for common divisors in Z.   To find the greatest common divisor of m and n (and hence all common divisors) we form the set

$$\mathfrak{I} = \{mx + ny \,|\, x, y \in Z\}$$

and look for the integer d for which

$$\mathfrak{I} = dZ.$$

Then

$$d = (m, n)$$

(Theorem 14 in Appendix 1).   Thus our search for π leads us to form

$$\mathfrak{D} = \{2x + (1 + \sqrt{-5})y \mid x, y \in \mathbf{A}(-5)\}.$$

Now we ask for the algebraic integer δ for which

$$\mathfrak{D} = \delta \mathbf{A}(-5).$$

Unfortunately, no such δ exists.   We must settle for less.   The subset $\mathfrak{D}$ of $\mathbf{A}(-5)$ is an ideal of $\mathbf{A}(-5)$ (in the ring theoretical sense).   Let us look for a context in which we can reasonably say that $\mathfrak{D}$ itself, rather than its "ideal" (but nonexistent) generator δ, is the greatest common divisor of 2 and $1 + \sqrt{-5}$.   Write

$$\mathfrak{D} = (2, 1 + \sqrt{-5}).$$

Note that

$$(2, 1 + \sqrt{-5}) = (2, 1 - \sqrt{-5}).$$

Consider the set $\mathbf{S}$ of ideals (again, in the ring theoretical sense) of $\mathbf{A}(-5)$. if $\mathfrak{A}$ and $\mathfrak{B}$ are in $\mathbf{S}$ define

$$\mathfrak{A}\mathfrak{B} = \left\{\sum_{i=j}^{n} a_i b_i \,\middle|\, a_i \in \mathfrak{A},\, b_i \in \mathfrak{B}\right\}.$$

The set $\mathbf{S}$ equipped with this multiplication is a semigroup in which all the definitions of Appendix 1 make sense.   The ideal $\mathbf{A}(-5)$ is the identity of $\mathbf{S}$; in $\mathbf{S}$ we can talk of prime ideals and divisibility.   The miracle is that $\mathbf{S}$ enjoys unique factorization.

To see how that helps us, let us see how the multiplicative structure of $\mathbf{A}(-5)$ is given by a part of $\mathbf{S}$.   If $\alpha \in \mathbf{A}(-5)$, write (α) for the principal ideal $\alpha\mathbf{A}(-5)$ consisting of all multiples of α.   Then α and α' are associates if and only if $(\alpha) = (\alpha')$, and $\alpha \mid \beta$ in $\mathbf{A}(-5)$ if and only if $(\alpha) \mid (\beta)$ in $\mathbf{S}$.   Thus if we wish to factor in $\mathbf{A}(-5)$, nothing is lost by passing to $\mathbf{S}$, and much is in fact gained.   We have found the "ideal" prime π which was to divide both 2 and $1 + \sqrt{-5}$.   In $\mathbf{S}$

$$(2, 1 + \sqrt{-5})(2, 1 - \sqrt{-5}) = (2, 1 + \sqrt{-5})^2 = (2)$$

and

$$(2, 1 + \sqrt{-5})(3, 1 + \sqrt{-5}) = (1 + \sqrt{-5})$$

so $(2, 1 + \sqrt{-5})$ *is* a common divisor of (2) and of $(1 + \sqrt{-5})$. In fact $(2, 1 + \sqrt{-5})$ is prime in **S**, and

$$(6) = (2, 1 + \sqrt{-5})^2(3, 1 + \sqrt{-5})(3, 1 - \sqrt{-5}) \tag{71}$$

is the unique factorization of (6) into prime ideals.   The expressions

$$(6) = (2)(3) = (1 + \sqrt{-5})(1 - \sqrt{-5})$$

result from collecting the factors in Eq. (71).   Our original naive faith that the source of the difficulty in Eq. (70) was that the factors on the right were not "really" prime leads to a new, beautiful, useful theory in which unique factorization is restored and with which we could renew our attack on

$$x^2 + 5y^2 = n$$

or more generally, on Eq. (69).

40. SUMS OF FOUR SQUARES

In this section we shall prove a theorem guessed by Fermat and first proved by Lagrange: Every positive integer is a sum of four squares.   The proof we give is due to Hurwitz.   It depends on the fact that enough of the machinery we have developed for Euclidean number fields works in a convenient non-commutative subring of the division ring of rational quaternions.

We wish now to introduce quaternions as rapidly as possible.   A more elegant approach will be found in Problem 41.44.   Let i, j, k be fixed symbols. Consider the set of formal sums

$$\mathbf{H} = \{a + bi + cj + dk \,|\, a, b, c, d \in \mathbf{Q}\}$$

where $\alpha = a + bi + cj + dk$ and $\alpha' = a' + b'i + c'j + d'k$ are equal in **H** if and only if $a = a'$, $b = b'$, $c = c'$, and $d = d'$.   Define

$$\alpha + \alpha' = (a + a') + (b + b')i + (c + c')j + (d + d')k$$

and

$$\alpha\alpha = a'' + b''i + c''j + d''k$$

where

$$a'' = aa' - bb' - cc' - dd'$$
$$b'' = ab' + ba' + cd' - dc'$$
$$c'' = ac' + ca' + db' - bd'$$
$$d'' = ad' + da' + bc' - cb'.$$

There is no need to remember these equations, for they are the result of formally expanding the product $\alpha\alpha'$ under the assumptions that both distributive laws hold and that

$$\left.\begin{aligned} ai &= ia \\ aj &= ja \\ ak &= ka \end{aligned}\right\} \text{ for all rational } a$$

$$i^2 = j^2 = k^2 = -1$$
$$ij = -ji = k$$
$$jk = -kj = i$$
$$ki = -ik = j. \tag{72}$$

We naturally regard $\mathbf{Q}$ as a subset of $\mathbf{H}$; $a \in R$ corresponds to $a + 0i + 0j + 0k \in \mathbf{H}$.

40.1 Theorem. The set $\mathbf{H}$ with these operations is a noncommutative ring with identity $1 = 1 + 0i + 0j + 0k$. The center of $\mathbf{H}$ is $\mathbf{Q}$.

The proof is left to the reader, who may tediously verify the associativity of multiplication and both distributive laws or consult Problem 41.44 instead.

40.2 Definition. The conjugate α^* and the norm $N(\alpha)$ of $\alpha \in \mathbf{H}$ are defined by

$$\alpha^* = a - bi - cj - dk$$

$$N(\alpha) = \alpha\alpha^* = a^2 + b^2 + c^2 + d^2. \tag{73}$$

(Compare Definition 28.3.)

40.3 Lemma. Conjugation is a ring antiautomorphism. That is,

$$(\alpha + \beta)^* = \alpha^* + \beta^* \tag{74}$$

$$(\alpha\beta)^* = \beta^*\alpha^*. \tag{75}$$

Moreover

$$(\alpha^*)^* = \alpha \tag{76}$$

and $\alpha = \alpha^*$ if and only if α is rational.

Proof. Equations (74) and (76) follow trivially from Definition 40.2. Equation (75) requires verification which we leave to the reader.

Finally, $\alpha = \alpha^*$ if and only if $b = c = d = 0$, so that the last assertion is trivial given our identification of **Q** with a subset of **H**.

Next we prove the analogue of Theorem 28.5.

40.4 Theorem.
(a) $N(\alpha\beta) = N(\alpha)N(\beta)$.
(b) $N(\alpha)$ is rational.
(c) $N(\alpha) = 0$ if and only if $\alpha = 0$.
(d) $N(\alpha) = \alpha^2$ if and only if α is rational.
(e) If $\alpha \neq 0$, then $N(\alpha)^{-1}\alpha^*$ is an inverse for α in **H**.

Proof. Parts (b), (c), and (d) are immediate consequences of Eq. (73). To prove (e) observe that $N(\alpha) = N(\alpha^*)$ and that $N(\alpha)$ commutes with α so that

$$\alpha(N(\alpha)^{-1}\alpha^*) = N(\alpha)^{-1}\alpha\alpha^* = 1$$

and

$$(N(\alpha^*)^{-1}\alpha^*) = N(\alpha^*)^{-1}N(\alpha^*) = 1.$$

Finally, (a) follows from

$$
\begin{aligned}
N(\alpha\beta) &= \alpha\beta(\alpha\beta)^* \\
&= \alpha\beta\beta^*\alpha^* \quad \text{(Eq. (75))} \\
&= \alpha N(\beta)\alpha^* \\
&= \alpha\alpha^* N(\beta) \quad \text{since } N(\beta) \text{ is rational and hence commutes with } \alpha^* \\
&= N(\alpha)N(\beta).
\end{aligned}
$$

40.5 Corollary. The quaternions are a division ring; that is, nonzero quaternions have multiplicative inverses.

Next we introduce a subring of **H** analogous to the rings of algebraic integers we have been studying.

40.6   Definition.

$\mathbf{D} = \{a + bi + cj + dk \mid 2a, 2b, 2c, 2d \text{ are integers of the same parity}\}$

We call quaternions in $\mathbf{D}$ integral.

40.7   Theorem.   The set D is a subring of $\mathbf{H}$; $\alpha \in D$ implies $\alpha + \alpha^*$ and $N(\alpha) \in \mathbf{Z}$.

Proof.   Let $\xi = \frac{1}{2}(1 + i + j + k)$.   Then it is easy to verify that

$$\mathbf{D} = \mathbf{Z}\xi + \mathbf{Z}i + \mathbf{Z}j + \mathbf{Z}k.$$

Thus $\mathbf{D}$ is closed under subtraction.

Since $\xi^2 = \xi - 1$, which is integral, an argument analogous to that in Theorem 29.7 shows $\mathbf{D}$ is closed under multiplication.

Clearly $\alpha + \alpha^* = 2a \in \mathbf{Z}$.   Moreover $4N(\alpha) = (2a)^2 + (2b)^2 + (2c)^2 + (2d)^2 \in \mathbf{Z}$ by assumption.   Since $2a \equiv 2b \equiv 2c \equiv 2d$ (2), we have $(2a)^2 \equiv (2b)^2 \equiv (2c)^2 \equiv (2d)^2$ (4), and hence $N(\alpha) \in \mathbf{Z}$.   The converse of these remarks is false; $\alpha = (1/3)i + (2/3)j + (2/3)k \notin \mathbf{D}$, but $\alpha + \alpha^* = 0 \in \mathbf{Z}$ and $\alpha\alpha^* = 1 \in \mathbf{Z}$. That is why we used 40.6 rather than the analogue of 29.1 to define integral quaternions.

Let $\mathbf{E}$ be the set of norms of integral quaternions.   Then $\mathbf{E} \subseteq \mathbf{Z}$.

40.8   Lemma. $b_1, b_2 \in \mathbf{E} \Rightarrow b_1 b_2 \in \mathbf{E}$.

Proof. $b_1 = N(\alpha_1)$ and $b_2 = N(\alpha_2)$ for $\alpha_1, \alpha_2 \in \mathbf{D}$.   Then

$$\alpha_1 \alpha_2 \in \mathbf{D}$$

so

$$b_1 b_2 = N(\alpha_1 \alpha_2) \in \mathbf{E}.$$

(Compare Lemma 30.3.)

40.9   Theorem.   The following are equivalent:
(a) n is the norm of an integral quaternion ($n \in \mathbf{E}$).
(b) $4n$ is a sum of four squares.
(c) n is a sum of four squares.

Proof.   Suppose (a) is true; say $n = N(\alpha)$ and $\alpha \in \mathbf{D}$.   Then

$$4n = (2a)^2 + (2b)^2 + (2c)^2 + (2d)^2$$

so (b) is true.

Suppose the even number $2m$ satisfies

$$2m = x^2 + y^2 + z^2 + w^2$$

for rational integers x, y, z, and w. Then these integers are either all
even, all odd, or just two are even (Problem 11.1). We may thus assume
$x \equiv y(2)$; $z \equiv w(2)$. Then

$$m = \left(\frac{x-y}{2}\right)^2 + \left(\frac{x+y}{2}\right)^2 + \left(\frac{z-w}{2}\right)^2 + \left(\frac{z+w}{2}\right)^2$$

so m is a sum of four integral squares. Two applications of these remarks
show (c) follows from (b).

If (c) is true, then $n = a^2 + b^2 + c^2 + d^2 = N(\alpha)$; $\alpha \in \mathbf{D}$ because a, b, c,
$d \in \mathbf{Z}$. (Compare Problem 6.2, Theorem 37.4, and Lemma 30.2.)

Thus we wish to show $\mathbf{E} = \mathbf{Z} \cup \{0\}$. In light of Lemma 40.8 it suffices
to show $p \in \mathbf{E}$ for every prime p. Our method will be to show that $\mathbf{D}$ is close
enough to being a Euclidean domain for us to apply the arguments in Defini-
tion 34.1 and Theorem 34.5.

As usual, $\mu \in \mathbf{D}$ is a *unit* when it is invertible in $\mathbf{D}$. The proof of Lemma
30.6 applies here to show that μ is a unit if and only if $N(\mu) = 1$. As usual,
the units form a group under multiplication, although the proof in Lemma 3
of Appendix 1 will not work; we argue instead as follows: If μ and v are units,
then $N(\mu v) = N(\mu)N(v) = 1$, so μv is one too.

40.10 Definition. The integral quaternion α is a *left divisor* of $\beta \in \mathbf{D}$
if and only if $\beta = \alpha\gamma$ for some $\gamma \in \mathbf{D}$. We say then that α left divides β and
that β is a right multiple of α and write $\alpha \mid \beta$. We need no symbol like $\alpha \mid_L \beta$
because we shall make no use of right divisors. The set of right multiples
of α is just the right ideal $\alpha\mathbf{D}$. A quaternion δ is a *greatest common left divisor*
of α and β if it left divides α and β and any γ which left divides α and β. We
say α and β are *left associates* when $\beta = \alpha\mu$ for some unit μ or, equivalently,
$\alpha \mid \beta$ and $\beta \mid \alpha$ (Theorem 5 of Appendix 1, in the proof of which no com-
mutativity is used). The quaternion π is a *left prime* if and only if its
only left divisors are units and its left associates.

40.11 Lemma. If $\alpha \neq 0$ and β are in $\mathbf{D}$, there is a $\tau \in \mathbf{D}$ such that
$N(\tau - \alpha^{-1}\beta) < 1$.

Proof. We modify the argument in Theorem 33.5. Let $\alpha^{-1}\beta = w + xi$
$+ yj + zk \in \mathbf{H}$. We can choose $a \in (\frac{1}{2})\mathbf{Z}$ such that $|a - w| \leq \frac{1}{4}$; then find

$b, c,$ and $d \in a + \mathbf{Z}$ such that $|b - x| \leq \frac{1}{2}$, $|c - y| \leq \frac{1}{2}$ and $|d - z| \leq \frac{1}{2}$. Let $\tau = a + bi + cj + dk \in \mathbf{D}$. Then

$$N(\tau - \alpha^{-1}\beta) = (a - w)^2 + (b - x)^2 + (c - y)^2 + (d - z)^2$$
$$\leq \tfrac{1}{16} + \tfrac{1}{4} + \tfrac{1}{4} + \tfrac{1}{4}$$
$$< 1.$$

40.12 Corollary. If $\alpha \neq 0$ and β are in $\mathbf{D}$, there are elements $\tau, \rho \in \mathbf{D}$ such that

$$\beta = \alpha\tau + \rho$$

and either $\rho = 0$ or $N(\rho) < N(\alpha)$.

Proof. Reread the proof of Lemma 33.3, substituting $\alpha^{-1}\beta$ for β/α.

40.13 Corollary. Let $\mathfrak{I} \neq \{0\}$ be a right ideal in $\mathbf{D}$. Then $\mathfrak{I} = \alpha\mathbf{D}$ for any $\alpha \in \mathfrak{I}$ whose norm is minimal among norms of elements of $\mathfrak{I}$.

Proof. See Theorem 10 in Appendix 1.

40.14 Theorem. Let α and β be nonzero integral quaternions. Then α and β have a greatest common left divisor δ, and there are integral quaternions ρ and σ such that

$$\delta = \alpha\rho + \beta\sigma.$$

Proof. See Theorem 14 of Appendix 1.

40.15 Lemma. Suppose the left prime π in $\mathbf{D}$ left divides the product $\alpha\beta$ and that $\pi\alpha = \alpha\pi$. Then π left divides α or β.

Proof. Suppose $\pi \nmid \beta$. Then the only common left divisors of π and β are units, so 1 is a greatest common left divisor of π and β and there are integral quaternions ρ and σ for which

$$1 = \pi\rho + \beta\sigma.$$

Then

$$\alpha = \alpha\pi\rho + \alpha\beta\sigma$$
$$= \pi\alpha\rho + \pi\gamma\sigma$$

where we have used the facts that π commutes with α and that $\pi \mid \alpha\beta$. Thus $\alpha = \pi(\alpha\rho + \gamma\sigma)$, so $\pi \mid \alpha$.

Next we prove an analogue of Lemma 26.3.

40.16 Lemma. Let p be an odd prime. Then there are integers x and y for which $1 + x^2 + y^2 \equiv 0(p)$.

Proof. Let $n = (p - 1)/2$ and let $r_1, \ldots, r_n$ be the quadratic residues of p in $\mathbf{Z}_p$ (Theorem 23.2). Set $r_0 = 0^2 = 0$. Then the two $n + 1$ element subsets $\{r_0, r_1, \ldots, r_n\}$ and $\{-1 - r_0, -1 - r_1, \ldots, -1 - r_n\}$ of the $p = 2n + 1$ element set $\mathbf{Z}_p$ must intersect. Thus $r_j = -1 - r_i$ for some i and j. If $x^2 \equiv r_j(p)$ and $y^2 \equiv r_i(p)$, then $1 + x^2 + y^2 \equiv 0(p)$. Note that when $p \equiv 1(4)$ we may take $y = 0$.

40.17 Theorem. An odd prime p is a sum of four squares.

Proof. Let p be such a prime. We wish to show p is not a left prime in $\mathbf{D}$. Use Lemma 40.17 to find integers x and y for which $p \mid 1 + x^2 + y^2$. Let $\gamma = 1 + xi + yj \in \mathbf{D}$; then $p \mid \gamma\gamma^*$ in $\mathbf{Z}$ and hence in $\mathbf{D}$. Clearly $p\gamma = \gamma p$, so if p were a left prime it would left divide γ or γ^* (Lemma 40.15), but neither γ/p nor

$$\frac{\gamma^*}{p} = \frac{1}{p} - \frac{x}{p}i - \frac{y}{p}j$$

is integral. Thus we can write $p = \alpha\beta$ where α is neither a unit nor a left associate of π; that is, neither α nor β is a unit. Then

$$p^2 = N(p) = N(\alpha)N(\beta).$$

Since $N(\alpha) > 1$ and $N(\beta) > 1$ we have $N(\alpha) = N(\beta) = p$; hence p is a sum of four squares (Theorem 40.9).

40.18 Theorem. Every positive integer is a sum of four squares.

Proof. The set of integers so representable is closed under multiplication (Theorem 40.9 and Lemma 40.8) and contains $2 = 1^2 + 1^2 + 0^2 + 0^2$ and every odd prime (Theorem 40.17).

41. PROBLEMS

41.1 Prove: If $m \neq n$ and both are square free, then $\mathbf{Q}(\sqrt{m}) \cap \mathbf{Q}(\sqrt{n}) = \mathbf{Q}$.

41.2 Show $\mathbf{Q}(\sqrt{m})$ has no field automorphisms other than the identity and conjugation.

41.3 Prove: If $m > 0$ then conjugation in $\mathbf{Q}(\sqrt{m})$ cannot be extended to be a continuous automorphism of $\mathbf{R}$.

41.4 Prove $\alpha \in \mathbf{Q}(\sqrt{m})$ is an algebraic integer if and only if it is a root of some monic polynomial in $\mathbf{Z}[x]$.

41.5 If $m > 0$ has a prime factor congruent to 3 modulo 4, then $A(m)$ has no improper unit. (See Lemma 30.7.) The converse is false; show $A(34)$ has no no improper unit.

41.6 Find the fundamental unit in $A(5)$, $A(6)$, and $A(7)$.

41.7 Prove: If $A(m)$ has an improper unit, then the fundamental unit μ_0 is improper, and the proper units are $\pm\mu_0^{2k}$.

41.8 The *fundamental solution* $\langle x_0, y_0 \rangle$ to Pell's equation

$$x^2 - my^2 = 1$$

is the one which minimizes $x + y\sqrt{m}$ for x and y positive. Show that if $m \not\equiv 1(4)$, $x_0 + y_0\sqrt{m}$ is either the fundamental unit of $A(m)$ or its square, depending on whether or not μ_0 is proper. When $m \equiv 1(4)$, $x_0 + y_0\sqrt{m} = \mu_0^k$ for some k. Show $k = 6$ when $m = 5$.

41.9 If p is a prime congruent to 1 modulo 4, then $A(p)$ has an improper unit. *Hint*: let $\langle x_0, y_0 \rangle$ be the fundamental solution to the Pell equation $x^2 - py^2 = 1$ (Problem 41.8). Show $x + y\sqrt{p}$ has a square root in $A(p)$ which is an improper unit.

41.10 Suppose $m \equiv 3(4)$, n is odd, and $\pm n \in B(m)$. Show that the minus sign is appropriate if and only if $n \equiv 3(4)$.

41.11* Learn how to solve Pell's equation using continued fractions.

41.12* Show that the semigroup of integers congruent to 1 modulo 4 does not enjoy unique factorization. (This example is due to Hilbert.)

41.13 Show $A(-6)$ and $A(10)$ are not unique factorization domains (compare Problem 27.9).

41.14 Let R be the subring of $\mathbf{Q}$ containing those rational numbers which can be written with odd denominators. Show that R is a unique factorization domain with just one prime (up to associates).

41.15 Find a unique factorization domain which contains just n primes.

41.16 Is $|N|$ a Euclidean norm for the subring $\mathbf{Z} + \mathbf{Z}\sqrt{-3}$ of $A(-3)$?

41.17 Prove Theorem 34.6. Factor 2 in $A(m)$ when $m = -11$, 5, 6, and 17.

41.18* Need $A(m)$ be a unique factorization domain when $B(m)$ enjoys unique factorization?

41.19 Show that -1 is an appropriate sign for n in Corollary 34.10 if and only if $-1 \in B(m)$, or α_j is odd for an odd number of primes p_j for which -1 is appropriate.

41.20* Let p and q be odd rational primes, and suppose $A(p)$ is a unique factorization domain. Prove $\left(\dfrac{p}{q}\right) = 1$ implies $\left(\dfrac{q}{p}\right) = 1$ unless $p \equiv q \equiv 3(4)$, in which case $\left(\dfrac{q}{p}\right) = -1$. That is, when $A(p)$ is a unique factorization domain, Theorem 34.6 implies half of the Quadratic Reciprocity Law. In fact with the ideal theory sketched in Section 39 we could drop the assumption above that $A(p)$ is a unique factorization domain and thus prove the Quadratic Reciprocity law by a method somewhat better motivated than the elementary one we used in Chapter 5. For details see Sommer (listed in the Bibliography).

The next five problems show that when $m < 0$, $A(m)$ is rarely a unique factorization domain.

41.21 Suppose $m < 0$ and $A(m)$ enjoys unique factorization. Let p be a prime $< |m|$ (if $m \not\equiv 1(4)$) or $< (|m|/4)$ (if $m \equiv 1(4)$). Show p is inertial in $A(m)$.

41.22 Prove: $m < -1$ and $A(m)$ a unique factorization domain imply $-m$ is prime.

41.23 Prove: $m < -7$ and $A(m)$ a unique factorization domain imply $-m$ is a prime congruent to 3 modulo 8. *Hint*: How does 2 factor in $A(m)$?

41.24 Prove: $m < -11$ and $A(m)$ a unique factorization domain imply $-m \equiv 19(24)$.

41.25* Extend the method of Problems 41.23 and 41.24 far enough to show $A(m)$ is not a unique factorization domain when $-200 < m < 0$ except possibly for the m listed in Table 2 (Section 33).

41.26 Suppose $m < -1$ and $A(m)$ enjoys unique factorization. Show that $x^2 + x + (1 - m)/4$ is a rational prime for $x = 0, 1, \ldots, ((1 - m)/4) - 2$. *Hint*: Look at $N(x + \xi)$. When $m = -163$, this yields the famous prime producing polynomial $x^2 + x + 41$.

The next five problems show that when $A(m)$ enjoys unique factorization it behaves very much like $\mathbf{Z}$. When $\alpha \in A(m)$ let $A(m)_\alpha$ be the quotient ring $A(m)/\alpha A(m)$. For some hints look ahead to Section 45.

41.27* Show $A(m)_\alpha$ has $|N(\alpha)|$ elements. Let $\Phi(\alpha)$ be the group of units in $A(m)_\alpha$ and $\varphi(\alpha)$ the order of $\Phi(\alpha)$. Show that the image of $\beta \in A(m)$ in $A(m)_\alpha$ is in $\Phi(\alpha)$ if and only if $(\alpha, \beta) = 1$. (Compare Definition 9.1 and Theorem 9.2.)

41.28* Prove φ is multiplicative.

41.29* Prove Euler's theorem for $A(m)_\alpha$. (Theorem 9.6.)

41.30 Suppose π is prime in $A(m)$. Show $A(m)_\pi$ is a field with $|N(\pi)|$ elements. Deduce that primes have primitive roots.

41.31* Generalize as much as you can of Chapters 2 and 4 to $A(m)$.

41.32* Read about the Lucas test for determining which Mersenne numbers are prime (Problem 6.18). See Hardy and Wright (listed in Bibliography), p. 223.

41.33 Count the number of relatively prime solutions to

$$x^2 + 2y^2 = n$$

in terms of the factorization of n in Z (see Example 35.5).

41.34 Euler knew how to factor an integer n when he could write it as a sum of squares two different ways. In our language his method is: If

$$n = a^2 + b^2 = c^2 + d^2$$

then one of $a \pm ib$ must share a Gaussian integral factor δ with one of $c \pm id$. Then δ can be found using the Euclidean algorithm in $A(-1)$ (Theorem 33.4 and Example 15 in Appendix 1). Then $\delta\bar{\delta}$ will be a proper factor of n.
  Factor $2501 = 50^2 + 1^2 = 49^2 + 10^2$ by this method.

41.35 Prove the following theorem of Niven's: The Gaussian integer $a + bi$ is a sum of two squares (of Gaussian integers) unless b is odd or $a \equiv b \equiv 2(4)$. (See Mordell, L. J.,"The Representation of a Gaussian Integer as a Sum of Two Squares," *Mathematics Magazine* **40**, (1967), 209.)

41.36 Solve the Diophantine equation $x^2 + y^2 = 21125$.

41.37 Find the least integer representable four ways as a sum of two squares.

41.38 Let a be the number of positive divisors d of n with $d \equiv 1(4)$ and b the number of positive divisors with $d \equiv 3(4)$. Show

$$x^2 + y^2 = n$$

has $\max(a - b, 0)$ solutions when they are properly counted.

41.39 State and prove a theorem analogous to Theorem 37.3 for each of the Diophantine equations

$$x^2 + 3y^2 = 4n \qquad \text{and} \qquad x^2 + 3y^2 = n.$$

41.40 Solve

$$x^2 - xy + y^2 = 1729$$
$$x^2 + 3y^2 = 4 \cdot 1729$$
$$x^2 + 3y^2 = 1729.$$

41.41 Theorem 37.4 is an accident peculiar to -3. That is, $A(m)$ a unique factorization domain, $m \equiv 1(4)$, and $n \in B(m)$ do not imply

$$x^2 - my^2 = n$$

has a solution. Find an example to show this. (Compare Problem 27.7.)

41.42 Prove the following generalization of Theorem 37.4. If n is odd, $m \equiv 1(8)$, and $A(m)$ enjoys unique factorization, then $x^2 - my^2 = n$ has a solution if and only if $n \in B(m)$.

41.43* Write the analogues of Section 35 through 38 for $m = -7, 3, 5$, and 6. Watch the signs for $m > 0$. (See Problems 27.10 and 41.19.)

41.44 Let $\mathcal{H}$ be the set of 2×2 matrices of the form

$$\begin{pmatrix} \alpha & \beta \\ -\bar{\beta} & \bar{\alpha} \end{pmatrix}$$

where $\alpha, \beta \in Q(i)$ (Section 28).
  (a) Show that $\mathcal{H}$ is a subring of the ring of all 2×2 matrices with coefficients in $Q(i)$ and that every nonzero element of $\mathcal{H}$ is invertible in $\mathcal{H}$.
  (b) Let I, J, and K be the elements

$$\begin{pmatrix} i & 0 \\ 0 & -i \end{pmatrix}, \quad \begin{pmatrix} 0 & 1 \\ -1 & 0 \end{pmatrix} \quad \text{and} \quad \begin{pmatrix} 0 & i \\ i & 0 \end{pmatrix}$$

of $\mathcal{H}$; when $a \in Q$, identify a with the diagonal matrix

$$\begin{pmatrix} a & 0 \\ 0 & a \end{pmatrix} \in \mathcal{H}.$$

Show that the map ψ given by $\psi(a + bi + cj + dk) = a + bI + cJ + dK$ is an isomorphism between the rational quaternions H (defined in Section 40) and $\mathcal{H}$. We could thus have avoided a tedious proof of Theorem 40.1 by defining "rational quaternion" as "element of $\mathcal{H}$."
  (c) Let α be an element of H. Prove: $\psi(\alpha^*) = \psi(\alpha)^*$, the conjugate of the transpose of the matrix $\psi(\alpha)$, and that $N(\alpha) = \det \psi(\alpha)$. Reprove Lemma 40.3 and Theorem 40.4 using these facts.

41.45 Show that every rational quaternion is a root of a quadratic polynomial with rational coefficients but that such polynomials can have more than two roots in H.

41.46 There are 24 units in D. Find them.

41.47 Can α and β be left associates in D without being right associates?

41.48 Show by example that Lemma 40.11 is false when D is replaced by its proper subring $Z + Zi + Zj + Zk$. (Compare Problem 41.16.)

41.49 Can we do without the hypothesis "p is prime" in Lemma 40.16?

41.50* Prove: If $m < 0$ and $m \neq -1$, -2, -3, -7, -11, then $A(m)$ is not Euclidean. *Hint*: Suppose E is a Euclidean norm for $A(m)$. Let $\sigma \in A(m)$ be a nonunit of minimal norm. Show that every $\alpha \in A(m)$ is congruent to 0 or ± 1 modulo σ and hence that $N(\sigma) \leq 3$ (Dubois, D. W., and Steger, A., "A Note on Division Algorithms in Imaginary Quadratic Number Fields," *Canadian Journal of Mathematics* 10, (1958), 285–286).

41.51* Show $|N|$ is a Euclidean norm for $A(6)$. (See Hardy and Wright, p. 214.)

7

The Fermat Conjecture

The most famous elementary conjecture in mathematics was left to us by Fermat in the margin of his copy of Bachet's edition of Diophantos's *Arithmetic*. He stated it (originally in Latin) as a theorem: "It is impossible to write a cube as the sum of two cubes, a fourth power as the sum of two fourth powers, and, in general, any power beyond the second as the sum of two similar powers. For this I have discovered a truly wonderful proof, but the margin is too small to contain it."

Fermat gave no proof and none has been found since, nor is a counter-example known. It is doubtful that Fermat knew a correct proof. We shall call his statement the Fermat conjecture rather than the more common Fermat's last theorem.

Conjecture. The Diophantine equation

$$x^n + y^n = z^n \tag{1}$$

has no nontrivial solutions when $n > 2$. The "solution" $1^3 + (-1)^3 = 0^3$ is to be regarded as trivial.

Since

$$x^{mn} = (x^m)^n$$

it would suffice to prove the conjecture whenever $n = 4$ or an odd prime. Fermat proved his conjecture for $n = 4$; Section 43 contains a proof which uses his methods. He probably knew, or thought he knew, a proof when $n = 3$ since he proposed it as a challenge to his contemporaries. Euler (1770) was the first to publish a proof in that case. We give one in Section 44, but not one Euler could have invented, for it is based on unique factorization in $\mathbf{A}(-3)$. In Section 45 we comment briefly on Kummer's generalization of that method, which is the starting point for most general assaults on the problem. The special case $n = 5$ was settled independently by Legendre and Dirichlet in about 1825. Lamé proved the Fermat conjecture for $n = 7$ in 1839.

42. PYTHAGOREAN TRIPLES

In order to prove the Fermat conjecture for $n = 4$ we must first find all the solutions to

$$x^2 + y^2 = z^2 \neq 0. \tag{2}$$

Equation (2) expresses the familiar relation among the sides of a right triangle, which motivates the next definition.

42.1 Definition. $\langle x, y, z \rangle$ is a *Pythagorean triple* if and only if Eq. (2) holds. We may assume x, y, and $z > 0$. The triple $\langle x, y, z \rangle$ is *primitive* if and only if $(x, y) = 1$.

Pythagorean triples exist; three familiar ones are

$$\langle 3, 4, 5 \rangle, \quad \langle 5, 12, 13 \rangle, \quad \text{and} \quad \langle 6, 8, 10 \rangle. \tag{3}$$

The first two of these are primitive.

If $\langle x, y, z \rangle$ is a Pythagorean triple, and $d = (x, y) > 1$, then

$$d^2 \mid (x^2 + y^2) = z^2$$

so $d \mid z$. Then

$$\left\langle \frac{x}{d}, \frac{y}{d}, \frac{z}{d} \right\rangle$$

is a primitive Pythagorean triple. Thus we will know all the Pythagorean triples once we know all the primitive ones. The observation which motivates the rest of this section is that

$$4uv + (u - v)^2 = (u + v^2). \tag{4}$$

Thus Eq. (4) yields a Pythagorean triple whenever uv is a square. The three triples in (3) correspond to

$$\langle u, v \rangle = \langle 4, 1 \rangle, \quad \langle 9, 4 \rangle, \quad \text{and} \quad \langle 8, 2 \rangle. \tag{5}$$

42.2 Lemma. If $\langle x, y, z \rangle$ is a primitive Pythagorean triple, then $x \not\equiv y(2)$.

Proof. If x and y were both even, $\langle x, y, z \rangle$ would not be primitive. If x and y were both odd, then $x^2 \equiv y^2 \equiv 1(4)$ would imply $z^2 \equiv 2(4)$, which is impossible (Problem 11.1).

Henceforth we shall always write our primitive Pythagorean triples so that x is even and y and z are odd.

42.3 Theorem. The equations

$$x = 2ab$$
$$y = a^2 - b^2$$
$$z = a^2 + b^2 \tag{6}$$

establish a one-to-one correspondence between primitive Pythagorean triples $\langle x, y, z \rangle$ and pairs $\langle a, b \rangle$ of relatively prime integers of opposite parity for which $a > b > 0$.

Proof. First suppose $a > b > 0$, $(a, b) = 1$, and $a \not\equiv b(2)$. Then Eq. (4) with $u = a^2$ and $v = b^2$ shows $\langle x, y, z \rangle$, defined by Eqs. (6), is a Pythagorean triple. Since $(a, b) = 1$ and $a \not\equiv b(2)$, x is even, y is odd, and $(x, y) = (2ab, a^2 - b^2) = 1$. Thus $\langle x, y, z \rangle$ is primitive.
Since

$$2a^2 = z + y \quad \text{and} \quad 2b^2 = z - y$$

$\langle x, y, z \rangle$ determines $\langle a, b \rangle$. Thus the theorem will be proved once we show Equations (6) construct every primitive Pythagorean triple.
Suppose $\langle x, y, z \rangle$ is such a triple. Then

$$x^2 = z^2 - y^2 = (z - y)(z + y).$$

Since x is even, and both z and y are odd, $(z - y)/2$ and $(z + y)/2$ are integers, and

$$\left(\frac{x}{2} \right)^2 = \left(\frac{z - y}{2} \right) \left(\frac{z + y}{2} \right). \tag{7}$$

Now $(z, y) = 1$, so $(z - y)/2$ and $(z + y)/2$ are relatively prime, and hence Eq. (7) implies each is a square, say

$$a^2 = \frac{z + y}{2}, \quad b^2 = \frac{z - y}{2}$$

where $a > b > 0$ and $(a, b) = 1$. Then Eqs. (6) are satisfied, and $a \not\equiv b(2)$, so we were done.

43. $x^4 + y^4 = z^4$; *THE METHOD OF DESCENT*

We can now quickly settle the Fermat conjecture when $n = 4$.

43.1   Theorem.   The Diophantine equation

$$x^4 + y^4 = z^2 \tag{8}$$

has no nontrivial solutions.   Thus, *a fortiori*, the Fermat conjecture is true for $n = 4$.

Proof.   By nontrivial we mean none of x, y, or z is zero.   Suppose $\langle x, y, z \rangle$ solves Eq. (8).   We may certainly assume x, y, and z positive and $(x, y) = 1$, by applying the argument following Definition 42.1 if necessary.   Thus

$$\langle x^2, y^2, z \rangle$$

is a primitive Pythagorean triple.   We may suppose x^2 and hence x even. Now apply Theorem 42.3.   There are relatively prime integers $a > b > 0$ of opposite parity such that

$$x^2 = 2ab \tag{9}$$

$$y^2 = a^2 - b^2 \tag{10}$$

$$z = a^2 + b^2. \tag{11}$$

Equation (9) and the fact that $(a, b) = 1$ imply that the odd member of the pair $\langle a, b \rangle$ is a square while the even member is twice a square.   Thus either

$$a = 2u^2, \quad b = v^2 \equiv 1 \ (2) \tag{12}$$

or

$$b = 2s^2, \quad a = (z')^2 \equiv 1 \ (2). \tag{13}$$

Let us show (12) is impossible.   It implies

$$4u^4 = a^2 = y^2 + b^2 = y^2 + v^4.$$

But both y and v^2 are odd, so

$$4u^4 = y^2 + v^4 \equiv 2(4)$$

which is absurd.   Thus (13) must be true.   Then

$$(z')^4 = a^2 = y^2 + 4s^4 = y^2 + (2s^2)^2 \tag{14}$$

so $\langle 2s^2, y, (z')^2 \rangle$ is a primitive Pythagorean triple.

Apply Theorem 42.3 again to find relatively prime integers t and w such that

$$2s^2 = 2tw \tag{15}$$

$$y = t^2 - w^2 \tag{16}$$

$$(z')^2 = t^2 + w^2. \tag{17}$$

Then Eq. (15) implies t and w are squares, so Eq. (17) implies $(z')^2$ is a sum of fourth powers, say

$$(x')^4 + (y')^4 = (z')^2.$$

Now observe that

$$(z')^2 = a < a^2 + b^2 = z \leq z^2.$$

We are ready to apply one of Fermat's favorite tools, a form of induction called the method of descent.   We have proved that from any solution $\langle x, y, z \rangle$ to Eq. (8) we could "descend" to one, $\langle x', y', z' \rangle$, for which $z' < z$. Thus the set of positive integers z for which Eq. (8) has a solution has no least member and so must be empty.   The theorem is proved.

44.   THE DIOPHANTINE EQUATION $x^3 + y^3 = z^3$

The crucial step in our identification of the Pythagorean triples (Theorem 42.3) and hence in our subsequent proof of the Fermat conjecture for $n = 4$ (Theorem 43.1) was the well known factorization of the difference of two squares:

$$x^2 - y^2 = (x - y)(x + y). \tag{18}$$

To study the Fermat conjecture for $n = 3$ we need the analogue of Eq. (18) for cubes, which begins

$$x^3 - y^3 = (x - y)(x^2 + xy + y^2). \tag{19}$$

We cannot factor further using integral coefficients but can when we allow coefficients from $\mathbf{A}(-3)$, which contains $\omega = (-1 + \sqrt{-3})/2$, a cube root of 1 (Lemma 29.4):

$$\omega^3 = 1. \tag{20}$$

Recall Eq. (12) of Chapter 6:

$$1 + \omega + \omega^2 = 0. \tag{21}$$

It follows that

$$x^3 - y^3 = (x - y)(x - \omega y)(x - \omega^2 y) \tag{22}$$

for all x and y in $\mathbf{A}(-3)$. In order to take advantage of Eq. (22) we shall prove more than we set out to. We shall show the Fermat conjecture for $n = 3$ is true for the algebraic integers $\mathbf{A}(-3)$.

In our discussion of Pythagorean triples and the Fermat conjecture for $n = 4$ we often used arguments based on parity, that is, on congruence modulo 2, or on congruence modulo 4. To study $x^3 + y^3 = z^3$, congruence modulo 3 is relevant. We define congruence in $\mathbf{A}(-3)$ in the obvious way.

44.1 Definition. If α, β and $\tau \in \mathbf{A}(-3)$, then

$$\alpha \equiv \beta \ (\tau)$$

if and only if

$$\tau \mid \alpha - \beta.$$

This definition and some of its consequences were explored in greater generality in Problems 41.27 through 41.31; we shall develop here only as much as we need to settle the Fermat conjecture for $n = 3$.

Write (τ) for the ideal of multiples of τ in the ring $\mathbf{A}(-3)$, and let

$$\pi_\tau \colon \mathbf{A}(-3) \to \mathbf{A}(-3)/(\tau)$$

be the natural ring homomorphism.

If 3 were prime in $\mathbf{A}(-3)$ we would study congruence modulo 3. However

$$-3 = (\sqrt{-3})^2$$

so 3 ramifies, and it is more useful to study instead congruence modulo $\sqrt{-3}$, the unique prime divisor of 3. We need some notation.

44.2 Definition. Let $\lambda = 1 - \omega = (3 - \sqrt{-3})/2 \in \mathbf{A}(-3)$.

44.3 Lemma. The integer λ is an associate of $\sqrt{-3}$ and hence is a prime divisor of 3 in $\mathbf{A}(-3)$. Thus $N(\lambda) = 3$. Moreover, for $\alpha, \beta \in \mathbf{A}(-3)$,

$$\alpha + \beta \equiv \alpha + \beta\omega \equiv \alpha + \beta\omega^2 \ (\lambda). \tag{23}$$

Proof. We just happen to have proved λ and $\sqrt{-3}$ associates in Section 32 (Eq. (40)). The last assertion of the lemma follows from the congruence

$$\omega = 1 - \lambda \equiv 1 \ (\lambda).$$

44.4 Corollary. Every element of $\mathbf{A}(-3)$ is congruent modulo λ to a rational integer.

Proof. Every element of $\mathbf{A}(-3)$ may be written in the form $a + b\omega$ with rational integers a and b (Lemma 29.6) and $a + b\omega \equiv a + b(\lambda)$ (Lemma 44.3).

44.5 Lemma. If c and d are rational integers, then

$$c \equiv d \ (\lambda) \quad \text{in} \quad \mathbf{A}(-3) \tag{24}$$

if and only if

$$c \equiv d \ (3) \quad \text{in} \quad \mathbf{Z}. \tag{25}$$

Proof. Since $\lambda \mid 3$ in $\mathbf{A}(-3)$, it is clear that congruence (25) implies congruence (24). Conversely, suppose (24) is true. Then

$$\lambda \mid c - d$$

so

$$3 = N(\lambda) \mid N(c - d) = (c - d)^2$$

which in turn implies

$$3 \mid c - d$$

since 3 is prime in **Z**.

Now we have collected enough information to identify the quotient ring $\mathbf{A}(-3)/(\lambda)$.

44.6 Theorem. The integers 0, 1, and -1 are a complete set of coset representatives of (λ) in $\mathbf{A}(-3)$; $\mathbf{A}(-3)/(\lambda)$ is isomorphic to the three element field $\mathbf{Z}_3$.

Proof. Lemma 44.5 shows that every $\alpha \in \mathbf{A}(-3)$ is in the same (λ)-coset as some rational integer and that every rational integer is congruent to 0, 1, or -1 modulo λ. $\mathbf{A}(-3)/(\lambda)$ thus has three elements which we shall naturally call 0, 1, and -1; it is clear that this labeling establishes an isomorphism with $\mathbf{Z}_3$.

We have often used the fact that if a is odd then $a^2 \equiv 1(8)$ (Problem 11.1). Here is its $\mathbf{A}(-3)$ analogue.

44.7 Lemma. $\alpha \equiv \pm 1(\lambda)$ implies

$$\alpha^3 \equiv \pm 1 \ (\lambda^4)$$

or, equivalently,

$$\alpha^3 \equiv \pm 1 \ (9).$$

Proof. The proof when $\alpha \equiv -1(\lambda)$ is similar to the one we are about to give when $\alpha \equiv 1(\lambda)$.
   We know

$$\alpha^3 - 1 = (\alpha - 1)(\alpha - \omega)(\alpha - \omega^2)$$

(Eq. (22)). Lemma 44.3 implies each factor on the right is congruent to $\alpha - 1$ and hence to 0 modulo λ, so $\alpha^3 \equiv 1(\lambda^3)$, but that is not quite good enough; we want λ^4. We know

$$\alpha = 1 + \beta\lambda$$

for some β. Since $\omega = 1 - \lambda$,

$$\omega^2 = 1 - 2\lambda + \lambda^2.$$

So

$$\alpha^3 - 1 = (\alpha - 1)(\alpha - \omega)(\alpha - \omega^2)$$
$$= \beta\lambda(\beta\lambda + \lambda)(\beta\lambda + 2\lambda - \lambda^2)$$
$$= \lambda^3\beta(\beta + 1)(\beta + 2 - \lambda).$$

Now Theorem 44.6 implies β, $\beta + 1$ and $\beta + 2 - \lambda$ are incongruent modulo λ and that therefore one of them is divisible by λ. Thus $\lambda^4 \mid \alpha^3 - 1$.

Since $\lambda \mid 3$ and 3 ramifies, λ^4 is an associate of 9, so the two assertions of the lemma are equivalent.

44.8   **Theorem** $\rho^3 + \sigma^3 + \tau^3 = 0$ has no solution with ρ, σ, and τ nonzero algebraic integers in $A(-3)$.

Before proceeding to the proof let us show that Theorem 44.8 implies the next theorem.

44.9   **Theorem.**   The Fermat conjecture is true for $n = 3$.

Proof. If $x^3 + y^3 = z^3$ in $\mathbf{Z}$ and $xyz \neq 0$, then

$$x^3 + y^3 + (-z)^3 = 0$$

in $A(-3)$, contradicting Theorem 44.8.

44.10   **Proof of Theorem 44.8.**   Suppose

$$\rho^3 + \sigma^3 + \tau^3 = 0 \tag{26}$$

and $\rho\sigma\tau \neq 0$ in $A(-3)$. We wish to show that this supposition leads to a contradiction. As in Sections 42 and 43 we can restrict our attention to primitive solutions. That is, we may assume $(\rho, \sigma) = 1$, for any common divisor of ρ and σ will divide τ and may be cancelled in Eq. (26). It then follows that

$$(\rho, \sigma) = (\sigma, \tau) = (\tau, \rho) = 1. \tag{27}$$

Note that we need the unique factorization of $A(-3)$ in order to assert the existence of these greatest common divisors.

The next step in our treatment of primitive Pythagorean triples was to show that just one of x, y, and z was even.

44.11   Lemma.   The prime λ divides just one of ρ, σ, and τ.

Proof.   Since $\langle \rho, \sigma, \tau \rangle$ is primitive, λ cannot divide more than one.   Suppose it divides none.   Then

$$\rho \equiv \pm 1 \ (\lambda), \ \sigma \equiv \pm 1 \ (\lambda), \ \tau \equiv \pm 1 \ (\lambda)$$

for one of the six possible choices of signs, and thus

$$0 = \rho^3 + \sigma^3 + \tau^3$$
$$\equiv \pm 1 \pm 1 \pm 1 \ (9) \tag{28}$$

(Lemma 44.7).   But (28) is false for all choices of signs, so the claim is true.

We may suppose $\lambda \mid \tau$, so that for some $n > 0$ and $\eta \in \mathbf{A}(-3)$,

$$\tau = \lambda^n \eta, \qquad \lambda \nmid \eta.$$

Then Lemma 44.11 and Eqs. (26) and (27) show

$$\rho^3 + \sigma^3 + \lambda^{3n} \eta^3 = 0, \tag{29}$$

$$(\rho, \sigma) = (\rho, \eta) = (\sigma, \eta) = 1, \tag{30}$$

and

$$\lambda \nmid \rho \sigma \eta \neq 0. \tag{31}$$

We wish to apply the method of descent to n; to do so we must prove slightly more than the impossibility of Eq. (29).   Suppose that for some unit μ of $\mathbf{A}(-3)$ and integer $n > 0$,

$$\rho^3 + \sigma^3 + \mu \lambda^{3n} \eta^3 = 0 \tag{32}$$

while (30) and (31) remain true.   That is, ρ, σ, η, and λ are mutually relatively prime.   Equation (29) is the special case $\mu = 1$ in Eq. (32).   We must have $n \geq 2$ in Eq. (32), for we have assumed $n > 0$ and $n = 1$ is ruled out by the impossibility of

$$\pm 1 \pm 1 \pm \mu \lambda^3 \equiv 0 \ (9). \tag{33}$$

To prove (33) impossible try each of the six units ± 1, $\pm \omega$, $\pm \omega^2$ in turn for μ (Theorem 30.8).

Now rearrange Eq. (32) and substitute in Eq. (22):

$$-\mu\lambda^{3n}\eta^3 = (\rho + \sigma)(\rho + \omega\sigma)(\rho + \omega^2\sigma). \tag{34}$$

The three factors on the right are congruent modulo λ (Lemma 44.3) and λ^3 divides their product, so λ divides each factor. Let

$$\beta_1 = \frac{\rho + \sigma}{\lambda}$$

$$\beta_2 = \frac{\rho + \omega\sigma}{\lambda}$$

$$\beta_3 = \frac{\rho + \omega^2\sigma}{\lambda} \in A(-3).$$

For future reference note that

$$\beta_1 + \omega\beta_2 + \omega^2\beta_3 = \frac{\rho}{\lambda}(1 + \omega + \omega^2) + \frac{\sigma}{\lambda}(1 + \omega^2 + \omega^4)$$

$$= 0 \tag{35}$$

by virtue of Eqs. (21) and (20), which implies $\omega^4 = \omega$.

Next we show β_1, β_2, and β_3 are mutually relatively prime. Observe that

$$\beta_1 - \beta_2 = \frac{(1 - \omega)\sigma}{\lambda} = \sigma$$

and

$$\omega\beta_1 - \beta_2 = \frac{(\omega - 1)\rho}{\lambda} = -\rho.$$

Thus any common factor of β_1 and β_2 divides both σ and ρ, so $(\rho, \sigma) = 1$ implies $(\beta_1, \beta_2) = 1$. That $(\beta_1, \beta_3) = (\beta_2, \beta_3) = 1$ follow from similar arguments.

Now we know

$$-\mu\lambda^{3n-3}\eta^3 = \beta_1\beta_2\beta_3 \tag{36}$$

where $n > 2$. Since the β_i are mutually relatively prime, Eq. (36) implies each is an associate of a cube in $A(-3)$, and λ^{3n-3} divides one of them.

We may assume that one is β_3 by replacing σ by $\omega\sigma$ or $\omega^2\sigma$ in Eqs. (32) and (34) if necessary. Thus there are units μ_1, μ_2 and μ_3 and mutually relatively prime algebraic integers ρ', σ', and η' each prime to λ such that

$$\beta_1 = \mu_1(\rho')^3$$
$$\beta_2 = \mu_2(\sigma')^3$$
$$\beta_3 = \mu_3 \lambda^{3n-3}(\eta')^3.$$

Substitute in Eq. (35):

$$\mu_1(\rho')^3 + \omega\mu_2(\sigma')^3 + \omega^2\mu_3 \lambda^{3n-3}(\eta')^3 = 0. \tag{37}$$

We are almost done. We need only dispose of two units to make Eq. (37) resemble Eq. (32). Multiply through by μ_1^{-1}. Then

$$(\rho')^3 + v(\sigma')^3 + \mu'\lambda^{3n-3}(\eta')^3 = 0 \tag{38}$$

for some units v and μ'. What can v be? We know ρ' and σ' are prime to λ, so

$$\rho' \equiv \pm 1(\lambda)$$
$$\sigma' \equiv \pm 1(\lambda).$$

Then reducing Eq. (38) modulo λ^3 yields

$$\pm v \pm 1 \equiv 0(\lambda^3). \tag{39}$$

Among the six possibilities ± 1, $\pm\omega$, $\pm\omega^2$ for v only the first two could fit Eq. (39), so $v = \pm 1$ and

$$(\rho')^3 + (\pm\sigma')^3 + \mu'\lambda^{3n-3}(\eta')^3 = 0.$$

Our descent from n to $n - 1$ is complete. There is no minimal $n > 0$ for which Eq. (32) can be solved, so it can never be solved.

45. CYCLOTOMIC FIELDS

Our proof of the Fermat conjecture for $n = 3$ rests on the fact that $\mathbf{A}(-3)$, unlike $\mathbf{Z}$, contains all three cube roots of 1, so that the factorization in Eq. (22) is possible. That suggests that the proper habitat of the Fermat conjecture for a particular prime p is a ring $\mathbf{W}(p)$ containing both $\mathbf{Z}$ and the p complex pth roots of 1.

The set G of pth roots of 1 is a finite subgroup of the group of nonzero complex numbers. Theorem 18.2 implies G is cyclic. The generators of G are called *primitive* pth roots of unity; let ξ be one. The pth roots of unity are the powers of ξ. When $p = 3$,

$$\xi = \omega = \frac{-1 + \sqrt{-3}}{2}$$

is a primitive cube root of 1. Let

$$\mathbf{Q}(\xi) = \{a_0 + a_1\xi + \cdots + a_{p-2}\xi^{p-2} \mid a_i \in \mathbf{Q}\} \tag{40}$$

and $\mathbf{W}(p)$ be the subset of $\mathbf{Q}(\xi)$ for which the rational a_i in (40) are all integers. Thus, for example, when $p = 3$, $\mathbf{Q}(\xi) = \mathbf{Q}(\omega)$, and $\mathbf{W}(3) = \mathbf{A}(-3)$.

$\mathbf{Q}(\xi)$ is a subfield of $\mathbf{C}$ (Compare Theorem 28.1) and $\mathbf{W}(p)$, whose elements are called *algebraic integers*, is a subring which contains precisely those members of $\mathbf{Q}(\xi)$ which are roots of monic polynomials in $\mathbf{Z}[x]$. (Compare Definition 29.1, Problem 41.4, and Theorem 29.7.)

In $\mathbf{W}(p)$

$$x^p - y^p = (x - y)(x - \xi y) \cdots (x - \xi^{p-1}y) \tag{41}$$

which generalizes Eq. (22).

When $\mathbf{W}(p)$ enjoys unique factorization, an argument like the one in Section 44 proves the Fermat conjecture for p in $\mathbf{W}(p)$ and hence *a fortiori* in $\mathbf{Z}$. The ring $\mathbf{W}(p)$ is a unique factorization domain for all odd primes p less than 23 but not when $p = 23$.

It was Dirichlet's observation that unique factorization might fail in $\mathbf{W}(p)$ which led Kummer to invent the theory of ideals whose rudiments we sketched for the quadratic case in Section 39. Using that theory he succeeded in proving the Fermat conjecture for a class of primes called *regular* primes, whose definition is too elaborate to be discussed here. The only irregular primes less than 100 are 37, 59, and 67. Kummer proved the Fermat conjecture for them by other methods. Unfortunately there are infinitely many irregular primes and may be only finitely many regular ones.

Fermat's conjecture thus remains a conjecture, though many special cases and partial results are known. (For details, consult Vandiver, H. S., "Fermat's Last Theorem," *American Mathematical Monthly* 53, (1946), 555–578.) It is important in the history of mathematics primarily because of the fruitfulness of inventions like the theory of ideals which were designed to prove it, failed, and yet served other ends.

46. PROBLEMS

46.1 Show that no right triangle with integral sides has an area which is a perfect square.

46.2 Find a natural one-to-one correspondence between right triangles whose legs are consecutive integers and the units of $A(2)$.

46.3* Theorem 42.3 implies that $z \in B(-1)$ when $\langle x, y, z \rangle$ is a primitive Pythagorean triple. Try to prove that directly (without using Theorem 42.3) in order to deduce Theorem 42.3 from the results in Section 36 and an analogue of Problem 41.33.

46.4 Use the methods of Section 42 to find all the solutions to each of the Diophantine equations

$$x^2 + 2y^2 = z^2 \qquad \text{and} \qquad x^2 + 3y^2 = z^2.$$

46.5* Prove that $x^4 + y^4 = z^2$ has no solutions in the Gaussian integers. (For a solution see Sommer, p. 188.)

46.6* $x^4 + y^4 = z^4$ has a solution in $A(m)$ only when $m = -7$. Prove that theorem, or read about it. (See Aigner, A., "Über die Möglichkeit von $x^4 + y^4 = z^4$ in quadratischen Körpern," *Jahresbericht Deutsch. Math. Verein.* **43**, (1934), 226–229.)

46.7 Show that the Diophantine equation

$$z^2 = x^4 - y^4$$

has no nontrivial solution.

1

Unique Factorization

In this appendix we shall state and prove some abstract algebraic theorems with important number theoretical applications (primarily in Sections 2, 12, 33, and 40).

Let R be an integral domain, that is, a commutative ring with identity $1 \neq 0$ and no 0 divisors. Let $R^* = R - \{0\}$. Let S be a subset of R^* which contains 1 and is closed under multiplication. Note that R^* itself has this property because R is an integral domain.

1. Definition. The element a *divides* b (written $a \mid b$) in S if and only if $b = ac$ for some $c \in S$. We say then too that a is a divisor of b and that b is a multiple of a. It is clear that $a \mid b$ and $b \mid c$ imply $a \mid c$.

2. Definition. The element $u \in S$ is a *unit* if it divides 1, or, equivalently, if there is a $v \in S$ such that $uv = 1$. The units are thus the invertible elements in S.

3. Lemma. The set U of units of S is a group under multiplication. Moreover, if $ab \in U$, then $a \in U$ and $b \in U$.

Proof. Since $1 \cdot 1 = 1$, $1 \in U$. If $uv = 1$ and $rs = 1$, then $(uv)(rs) = ur(vs)$ $= 1$, so $ur \in U$. If $uv = 1$, then v is a multiplicative inverse for u.

If $ab \in U$, then for some $c \in S$, $b(ac) = a(bc) = (ab)c = 1$. Hence a and b are units.

4. Definition. If $a \mid b$ and $b \mid a$, a and b are *associates*. When that happens write $a \sim b$. Since $1 \mid a$ for all a, the units are just the associates of 1.

5. Theorem. There is a unit u such that $a = bu$ if and only if a and b are associates. The relation $\sim$ is an equivalence relation.

Proof. Suppose $a \sim b$. Then there are elements c, $d \in S$ such that $a = bc$ and $b = ad$. Then $a = adc$, so $dc = 1$ (because S is a subset of an integral domain), and hence c and d are units. Conversely if $a = bu$, then $b \mid a$. But $u \in U$, so $u^{-1} \in S$, and $b = au^{-1}$. Thus $a \mid b$ as well, and $a \sim b$.

The relation $\sim$ is clearly reflexive ($a \sim a$ because $a \mid a$) and symmetric ($a \sim b$ implies $b \sim a$). To prove it is transitive suppose $a \sim b$ and $b \sim c$. Then there are units u and v such that $a = bu$ and $b = cv$. Then $a = cvu$ implies $a \sim c$ since vu is a unit.

6. Definition. A nonunit $p \in S$ is *prime* if and only if its only divisors are units and its associates. If p is prime, so are all its associates. If p and q are prime and $p \mid q$, then p and q are associates.

The subset S is a *factorization semigroup* when every nonunit in S can be written as a finite product of primes; S is a *unique factorization semigroup* when it is a factorization semigroup such that whenever

$$up_1 \cdots p_r = vq_1 \cdots q_k \tag{1}$$

in S and the p_i and q_j are primes and u and v are units, then $r = k$ and there is a permutation π of $\{1, \ldots, r\}$ such that $p_{\pi(i)}$ and q_i are associates for $i = 1, \ldots, r$. We then sometimes say S *enjoys unique factorization*. The integral domain R is a *unique factorization domain* (UFD) when R^* enjoys unique factorization.

7. Theorem. A factorization semigroup S enjoys unique factorization if and only if whenever a prime in S divides a product it divides one of the factors.

Proof. Suppose S enjoys unique factorization, p is prime in S, and $p \mid ab$. Let

$$ab = pc. \tag{2}$$

Since p is not a unit, Lemma 3 implies a and b are not both units. Write the nonunits among a, b, c as products of primes and consider Eq. (2). The primes on the right are a permutation of associates of those on the left, so p is an associate of a prime divisor of a or of b.   Hence $p \mid a$ or $p \mid b$.

Conversely, suppose that when a prime divides a product it divides one of the factors, and suppose

$$up_1 \cdots p_r = vq_1 \cdots q_k \tag{3}$$

where the p_i and q_j are primes and u and v are units.   We may assume $r \leq k$.   If $r = 0$, that is, if the left-hand side of (3) is a unit, then so is the right-hand side (Lemma 3), so $k = 0 = r$.   Suppose the theorem true for products of fewer than r_0 primes.   That is, suppose that when $r < r_0$ the truth of Eq. (3) implies $k = r$ and that the primes $p_1, \ldots, p_r$ are associates of the primes $q_1, \ldots, q_k$ in some order.   Examine (3) when the left member has $r_0 > 1$ prime factors.   Since $p_{r_0} \mid vq_1 \cdots q_k$ and is prime, it divides one of the factors; p_{r_0} does not divide v, so $p_{r_0} \mid q_i$ for some i.   We may assume $i = k$. Then $p_{r_0} = wq_k$ for some unit w, and

$$uwp_1 \cdots p_{r_0-1} = vq_1 \cdots q_{k-1}. \tag{4}$$

The induction hypothesis then implies $r_0 - 1 = k - 1$ and that the remaining primes p_i and q_j are associates in some order.   Thus the theorem is proved.

We now proceed to show that an important class of domains are unique factorization domains.

8.   Definition.   A function $E : R^* \rightarrow Z^+$ (the positive integers) is a *Euclidean norm* if it satisfies
 (a) $E(ab) = E(a)E(b)$.
 (b)   Given $a \neq 0$ and $b \in R$ there are elements q and $r \in R$ such that

$$b = aq + r$$

and either $r = 0$ or $E(r) < E(a)$.
The domain R is a *Euclidean domain* if it possesses a Euclidean norm.

9.   Examples.   The integers are a Euclidean domain since $E(a) = |a|$ is a Euclidean norm (Lemma 2.2).
 When F is a field, the domain $F[x]$ of polynomials with coefficients in F is a Euclidean domain.   Each nonzero polynomial a has a nonnegative degree d; set $E(a) = 2^d$.
 More examples can be found in Section 33.   A noncommutative ring which resembles a Euclidean domain is discussed in Section 40.

The next theorem says that ideals in R are principal. It will apply to right ideals in the noncummutative example mentioned above.

10. Theorem.   Let $\Im \neq \{0\}$ be a (right) ideal in the Euclidean domain R. Then

$$\Im = aR = \{ar \mid r \in R\}$$

for any $a \in \Im$ whose norm $E(a)$ is minimal among norms of elements of $\Im$.

Proof. Since the set of norms of nonzero elements of $\Im$ is a nonempty subset of Z^+, it has a least element n_0. Suppose $a \in \Im$ and $E(a) = n_0$. Clearly $aR \subseteq \Im$. Suppose $b \in \Im$; we wish to prove $b \in aR$. Find q and r satisfying (b) in Definition 8. Since $r = b - aq \in R$, we cannot have $E(r) < E(a)$, so $r = 0$. Then $b = aq \in aR$, so $\Im = aR$.

11. Corollary.   The element u is a unit of R if and only if $E(u) = 1$.

Proof. If u is a unit, then for some $v \in R$, $uv = 1$ so

$$E(u)E(v) = E(1) = E(1^2) = E(1)^2.$$

Thus $E(1) = E(u) = 1$.

Conversely, if $E(u) = 1$, its norm is minimal because all norms are positive. Then Theorem 10 applied to the ideal $\Im = R$ shows $uR = R$. Hence $1 \in uR$, so $1 = uv$ for some $v \in R$ and u is a unit.

12. Corollary.   A Euclidean domain is a factorization domain.

Proof. We proceed by induction on $E(a)$. If $E(a) = 1$, then a is a unit and there is nothing to prove. Suppose every nonunit $r \in R^*$ for which $E(r) < n$ is a finite product of primes, and suppose $E(a) = n \geq 2$. If a is prime, it is a product of primes. If a is not prime, we can write $a = bc$ where neither b nor c is a unit. Then $E(b) > 1$ and $E(c) > 1$. Since $n = E(a) = E(bc)$ $= E(b)E(c)$, both $E(b) < n$ and $E(c) < n$. Thus b and c and hence a can be written as products of primes.

13. Definition.   An element d is a *common divisor* of a and b if $d \mid a$ and $d \mid b$; d is a *greatest common divisor* (g.c.d.) if it is a common divisor and a multiple of every common divisor. Any two greatest common divisors are associates; any associate of a g.c.d. is again one.

14. Theorem. Let a and b be nonzero elements of the Euclidean domain R. Then a and b have a greatest common divisor d, and there are elements $r, s \in R$ such that

$$d = ar + bs.$$

Proof. Let $\mathfrak{D} = \{ax + by \mid x, y \in R\}$. Then $\mathfrak{D}$ is a nonzero (right) ideal of R, so $\mathfrak{D} = dR$ for some $d = ar + bs \in \mathfrak{D}$ (Theorem 10). Since $a = a \cdot 1 + b \cdot 0 \in dR$, $d \mid a$. Similarly, $d \mid b$. If c divides both a and b, so that $a = ca'$ and $b = cb'$, then $d = c(a'r + b's)$, so $c \mid d$. Thus d is a g.c.d.

15. Example. Theorem 14 merely asserts the existence of d, r, and s. We shall illustrate below the famous *Euclidean algorithm* with which we can compute them in a predictable number of steps. Suppose $a = 102$ and $b = 258$ in **Z**. Apply (b) of Definition 8 to write

$$258 = 102 \cdot 2 + 54. \tag{5}$$

It follows from Eq. (5) that any common divisor of 258 and 102 is one of 102 and 54 and conversely, so the problem has been reduced to the computation of the g.c.d. of the latter, smaller pair. The same method shrinks the numbers again:

$$102 = 1 \cdot 54 + 48$$
$$= 2 \cdot 54 - 6. \tag{6}$$

The second equation in (6) is more convenient since $E(-6) = |6| < E(48)$. Any g.c.d. of -6 and 54 is one of 54 and 102 and hence one of 102 and 258. Since

$$54 = (-9)(-6) \tag{7}$$

-6 is a g.c.d. of 102 and 258. To solve

$$-6 = 258r + 102s$$

work backwards.

$$-6 = 102 - 2 \cdot 54 \quad \text{(Eq. (6))}$$
$$= 102 - 2(258 - 2 \cdot 102) \quad \text{(Eq. (5))}$$
$$= 258(-2) + 102 \cdot 5. \tag{8}$$

This algorithm clearly works in any Euclidean domain and produces the analogue of Eq. (8) and hence a g.c.d. of a and b after at most $\min\{E(a), E(b)\}$ applications of (b) in Definition 8.

16. Theorem. A Euclidean domain R is a unique factorization domain.
Proof. In light of Theorem 7 and Corollary 12 we need only prove that when a prime p in R divides a product ab it divides a or b. Suppose p does not divide b. Then the only common divisors of p and b are units, so 1 is a g.c.d. of p and b. Theorem 14 implies there are elements $r, s \in R$ for which

$$1 = pr + bs.$$

Then

$$a = apr + abs. \tag{9}$$

Since $p \,|\, apr$ and, by assumption $p \,|\, ab$, Eq. (9) implies $p \,|\, a$.

We close this appendix with two important observations about the Euclidean domain $F[x]$ of polynomials with coefficients in the field F.

17. Theorem. The element a of F is a root of $f(x) \in F[x]$ if and only if $(x - a) \,|\, f(x)$.

Proof. If $(x - a) \,|\, f(x)$, then $f(x) = (x - a)g(x)$ for some polynomial g, so $f(a) = (a - a)g(a) = 0$, and a is a root of f.
    Conversely, suppose $f(a) = 0$. Write

$$f(x) = (x - a)q(x) + r(x)$$

where r is a polynomial of degree less than the degree of $x - a$, which is 1. That is, r is a constant polynomial. Then $0 = f(a) = (a - a)q(a) + r$, so $f(x) = (x - a)q(x)$ and $(x - a) \,|\, f(x)$.

18. Corollary. A polynomial of degree n with coefficients in a field has at most n roots.

Proof. Let $a_1, \ldots, a_r$ be the roots of $f(x) \in F[x]$. Then the polynomials $x - a_1, \ldots, x - a_r$ are different prime divisors of f. Hence $(x - a_1) \cdots (x - a_r)$, which is of degree r, divides f, which is of degree n. Thus $r \leq n$.

2

Primitive Roots

n	nth prime p	Least positive primitive root for p	*Means 10 is a primitive root for p
1	2		
2	3	2	
3	5	2	
4	7	3	*
5	11	2	
6	13	2	
7	17	3	*
8	19	2	*
9	23	5	*
10	29	2	*
11	31	3	
12	37	2	
13	41	6	
14	43	3	
15	47	5	*
16	53	2	
17	59	2	*
18	61	2	*

n	nth prime p	Least positive primitive root for p	*Means 10 is a primitive root for p
19	67	2	
20	71	7	
21	73	5	
22	79	3	
23	83	2	
24	89	3	
25	97	5	*
26	101	2	
27	103	5	
28	107	2	
29	109	6	*
30	113	3	*
31	127	3	
32	131	2	*
33	137	3	
34	139	2	
35	149	2	*
36	151	6	
37	157	5	
38	163	2	
39	167	5	*
40	173	2	
41	179	2	*
42	181	2	*
43	191	19	
44	193	5	*
45	197	2	
46	199	3	
47	211	2	
48	223	3	*
49	227	2	
50	229	6	*
51	233	3	*
52	239	7	
53	241	7	
54	251	6	
55	257	3	*
56	263	5	*
57	269	2	*
58	271	6	
59	277	5	
60	281	3	
61	283	3	
62	293	2	
63	307	5	
64	311	17	

n	nth prime p	Least positive primitive root for p	*Means 10 is a primitive root for p
65	313	10	*
66	317	2	
67	331	3	
68	337	10	*
69	347	2	
70	349	2	
71	353	3	
72	359	7	
73	367	6	*
74	373	2	
75	379	2	*
76	383	5	*
77	389	2	*
78	397	5	
79	401	3	
80	409	21	
81	419	2	*
82	421	2	
83	431	7	
84	433	5	*
85	439	15	
86	443	2	
87	449	3	
88	457	13	
89	461	2	*
90	463	3	
91	467	2	
92	479	13	
93	487	3	*
94	491	2	*
95	499	7	*
96	503	5	*
97	509	2	*
98	521	3	
99	523	2	
100	541	2	*

3

Indices for $\Phi(315)$

In this appendix we shall work through Theorems 21.1 and 17.10 explicitly for the special case $n = 315 = 3^2 \cdot 5 \cdot 7$. This exercise should help convince the reader that those theorems are constructive and do not merely show indirectly the existence of an isomorphism. Then we shall answer some questions in $\Phi(315)$ using the index calculus discussed in the last paragraph of Section 17.

The prime power factors of 315 are 9, 5, and 7. A primitive root for 9 and 5 is 2; 3 is one for 7. We can thus obtain generators for $\Phi(315)$ by solving the three sets of simultaneous congruences

$$g_9 \equiv 2\ (9),\ 1\ (5),\ \text{and}\ \ 1\ (7)$$

$$g_5 \equiv 1\ (9),\ 2\ (5),\ \text{and}\ \ 1\ (7)$$

$$g_7 \equiv 1\ (9),\ 1\ (5),\ \text{and}\ \ 3\ (7).$$

The first of these sets we solved in Example 8.2, where we found

$$g_9 = 281.$$

The second and third sets are in Problem 11.6. The answers are

$$g_5 = 127$$
$$g_7 = 136.$$

Then

$$\text{ind } x = \langle a, b, c \rangle$$

means

$$x \equiv 281^a 127^b 136^c \ (315).$$

Since $\varphi(9) = \varphi(7) = 6$, and $\varphi(5) = 4$,

$$\langle a, b, c \rangle \in \mathbf{Z}_6 \times \mathbf{Z}_4 \times \mathbf{Z}_6 = K.$$

The next computations use the table of indices and antiindices at the end of this Appendix.

To what exponent d does 256 belong modulo 315? Since

$$\text{ind } (256) = \langle 2, 0, 4 \rangle$$

we must compute the order of $\langle 2, 0, 4 \rangle$ in K. Lemmas 17.2 and 17.7 imply

$$d = \text{l.c.m.} \left\{ \frac{6}{(2, 6)}, \frac{4}{(0, 4)}, \frac{6}{(4, 6)} \right\}$$

$$= \text{l.c.m.}\{3, 1, 3\}$$

$$= 3.$$

Next we solve the congruence

$$x^3 \equiv 244 \ (315). \tag{1}$$

Set

$$\text{ind } x = \langle a, b, c \rangle.$$

Since

$$\text{ind } (244) = \langle 0, 2, 3 \rangle \tag{2}$$

(1) is equivalent to

$$3\langle a, b, c \rangle = \langle 0, 2, 3 \rangle$$

in K and thus to

$$3a \equiv 0 \ (6)$$

$$3b \equiv 2 \ (4)$$

$$3c \equiv 3 \ (6).$$

These entail $a = 0$, 2, or 4; $b = 2$; and $c = 1$, 3 or 5. There are thus 9 solutions:

ind x	x
$\langle 0, 2, 1 \rangle$	199
$\langle 0, 2, 3 \rangle$	244
$\langle 0, 2, 5 \rangle$	19
$\langle 2, 2, 1 \rangle$	94
$\langle 2, 2, 3 \rangle$	139
$\langle 2, 2, 5 \rangle$	229
$\langle 4, 2, 1 \rangle$	304
$\langle 4, 2, 3 \rangle$	34
$\langle 4, 2, 5 \rangle$	124

The reader is invited to ask and answer similar questions.

It is instructive to write a computer program or flow chart which will produce for a given integer n tables like those that follow. Organizing the arithmetic so that computer time (or one's own time on a desk calculator) is used efficiently will lead to a still better understanding of Theorems 21.1 and 17.10.

Table of Indices for $\Phi(315)$

ind $x = a, b, c$ means $x \equiv 281^a \ 127^b \ 136^c \ (315)$.

x	ind x		x	ind x
1	0, 0, 0		23	5, 3, 2
2	1, 1, 2		26	3, 0, 5
4	2, 2, 4		29	1, 2, 0
8	3, 3, 0		31	2, 0, 1
11	1, 0, 4		32	5, 1, 4
13	2, 3, 3		34	4, 2, 3
16	4, 0, 2		37	0, 1, 2
17	3, 1, 1		38	1, 3, 1
19	0, 2, 5		41	5, 0, 3
22	2, 1, 0		43	4, 3, 0

x	ind x		x	ind x
44	3, 2, 2		149	5, 2, 2
46	0, 0, 4		151	4, 0, 4
47	1, 1, 5		152	3, 1, 5
52	4, 1, 1		157	2, 1, 1
53	3, 3, 4		158	5, 3, 4
58	2, 3, 2		163	0, 3, 2
59	5, 2, 1		164	1, 2, 1
61	4, 0, 5		166	2, 0, 5
62	3, 1, 3		167	5, 1, 3
64	0, 2, 0		169	4, 2, 0
67	2, 1, 4		172	0, 1, 4
68	5, 3, 5		173	1, 3, 5
71	3, 0, 0		176	5, 0, 0
73	0, 3, 1		178	4, 3, 1
74	1, 2, 4		179	3, 2, 4
76	2, 0, 3		181	0, 0, 3
79	4, 2, 2		184	2, 2, 2
82	0, 1, 5		187	4, 1, 5
83	1, 3, 3		188	3, 3, 3
86	5, 0, 2		191	1, 0, 2
88	4, 3, 4		193	2, 3, 4
89	3, 2, 5		194	5, 2, 5
92	1, 1, 0		197	3, 1, 0
94	2, 2, 1		199	0, 2, 1
97	4, 1, 3		202	2, 1, 3
101	1, 0, 1		206	3, 0, 1
103	2, 3, 5		208	0, 3, 5
104	5, 2, 3		209	1, 2, 3
106	4, 0, 0		211	2, 0, 0
107	3, 1, 2		212	5, 1, 2
109	0, 2, 4		214	4, 2, 4
113	5, 3, 0		218	1, 3, 0
116	3, 0, 4		221	5, 0, 4
118	0, 3, 3		223	4, 3, 3
121	2, 0, 2		226	0, 0, 2
122	5, 1, 1		227	1, 1, 1
124	4, 2, 5		229	2, 2, 5
127	0, 1, 0		232	4, 1, 0
128	1, 3, 2		233	3, 3, 2
131	5, 0, 5		236	1, 0, 5
134	3, 2, 0		239	5, 2, 0
136	0, 0, 1		241	4, 0, 1
137	1, 1, 4		242	3, 1, 4
139	2, 2, 3		244	0, 2, 3
142	4, 1, 2		247	2, 1, 2
143	3, 3, 1		248	5, 3, 1
146	1, 0, 3		251	3, 0, 3
148	2, 3, 0		253	0, 3, 0

x	ind x		x	ind x
254	1, 2, 2		284	5, 2, 4
256	2, 0, 4		286	4, 0, 3
257	5, 1, 5		289	0, 2, 2
262	0, 1, 1		292	2, 1, 5
263	1, 3, 4		293	5, 3, 3
268	4, 3, 2		296	3, 0, 2
269	3, 2, 1		298	0, 3, 4
271	0, 0, 5		299	1, 2, 5
272	1, 1, 3		302	5, 1, 0
274	2, 2, 0		304	4, 2, 1
277	4, 1, 4		307	0, 1, 3
278	3, 3, 5		311	5, 0, 1
281	1, 0, 0		313	4, 3, 5
283	2, 3, 1		314	3, 2, 3

ind x	x		ind x	x
0, 0, 0	1		1, 1, 0	92
0, 0, 1	136		1, 1, 1	227
0, 0, 2	226		1, 1, 2	2
0, 0, 3	181		1, 1, 3	272
0, 0, 4	46		1, 1, 4	137
0, 0, 5	271		1, 1, 5	47
0, 1, 0	127		1, 2, 0	29
0, 1, 1	262		1, 2, 1	164
0, 1, 2	37		1, 2, 2	254
0, 1, 3	307		1, 2, 3	209
0, 1, 4	172		1, 2, 4	74
0, 1, 5	82		1, 2, 5	299
0, 2, 0	64		1, 3, 0	218
0, 2, 1	199		1, 3, 1	38
0, 2, 2	289		1, 3, 2	128
0, 2, 3	244		1, 3, 3	83
0, 2, 4	109		1, 3, 4	263
0, 2, 5	19		1, 3, 5	173
0, 3, 0	253		2, 0, 0	211
0, 3, 1	73		2, 0, 1	31
0, 3, 2	163		2, 0, 2	121
0, 3, 3	118		2, 0, 3	76
0, 3, 4	298		2, 0, 4	256
0, 3, 5	208		2, 0, 5	166
1, 0, 0	281		2, 1, 0	22
1, 0, 1	101		2, 1, 1	157
1, 0, 2	191		2, 1, 2	247
1, 0, 3	146		2, 1, 3	202
1, 0, 4	11		2, 1, 4	67
1, 0, 5	236		2, 1, 5	292

ind x	x	ind x	x
2, 2, 0	274	4, 1, 0	232
2, 2, 1	94	4, 1, 1	52
2, 2, 2	184	4, 1, 2	142
2, 2, 3	139	4, 1, 3	97
2, 2, 4	4	4, 1, 4	277
2, 2, 5	229	4, 1, 5	187
2, 3, 0	148	4, 2, 0	169
2, 3, 1	283	4, 2, 1	304
2, 3, 2	58	4, 2, 2	79
2, 3, 3	13	4, 2, 3	34
2, 3, 4	193	4, 2, 4	214
2, 3, 5	103	4, 2, 5	124
3, 0, 0	71	4, 3, 0	43
3, 0, 1	206	4, 3, 1	178
3, 0, 2	296	4, 3, 2	268
3, 0, 3	251	4, 3, 3	223
3, 0, 4	116	4, 3, 4	88
3, 0, 5	26	4, 3, 5	313
3, 1, 0	197	5, 0, 0	176
3, 1, 1	17	5, 0, 1	311
3, 1, 2	107	5, 0, 2	86
3, 1, 3	62	5, 0, 3	41
3, 1, 4	242	5, 0, 4	221
3, 1, 5	152	5, 0, 5	131
3, 2, 0	134	5, 1, 0	302
3, 2, 1	269	5, 1, 1	122
3, 2, 2	44	5, 1, 2	212
3, 2, 3	314	5, 1, 3	167
3, 2, 4	179	5, 1, 4	32
3, 2, 5	89	5, 1, 5	257
3, 3, 0	8	5, 2, 0	239
3, 3, 1	143	5, 2, 1	59
3, 3, 2	233	5, 2, 2	149
3, 3, 3	188	5, 2, 3	104
3, 3, 4	53	5, 2, 4	284
3, 3, 5	278	5, 2, 5	194
4, 0, 0	106	5, 3, 0	113
4, 0, 1	241	5, 3, 1	248
4, 0, 2	16	5, 3, 2	23
4, 0, 3	286	5, 3, 3	293
4, 0, 4	151	5, 3, 4	158
4, 0, 5	61	5, 3, 5	68

4

Fundamental Units in

Real Quadratic Number Fields

When m is square free, μ_0 is the fundamental unit. $\langle x, y \rangle$ is the fundamental solution to the Pell equation

$$x^2 - my^2 = D.$$

The value of D appears in the last column; $D = -1$ implies μ_0 is improper. When m is square free and $m \not\equiv 1(4)$, $\mu_0 = x + y\sqrt{m}$.

m	μ_0	x	y	D
2	$1 + \sqrt{2}$			-1
3	$2 + \sqrt{3}$			1
5	$\frac{1}{2}(1 + \sqrt{5})$	2	1	-1
6	$5 + 2\sqrt{6}$			1
7	$8 + 3\sqrt{7}$			1

m	μ_0	x	y	D
8		3	1	1
10	$3+\sqrt{10}$			-1
11	$10+3\sqrt{11}$			1
12		7	2	1
13	$\frac{1}{2}(3+\sqrt{13})$	18	5	-1
14	$15+4\sqrt{14}$			1
15	$4+\sqrt{15}$			1
17	$4+\sqrt{17}$	4	1	-1
18		17	4	1
19	$170+39\sqrt{19}$			1
20		9	2	1
21	$\frac{1}{2}(5+\sqrt{21})$	55	12	1
22	$197+42\sqrt{22}$			1
23	$24+5\sqrt{23}$			1
24		5	1	1
26	$5+\sqrt{26}$			-1
27		26	5	1
28		127	24	1
29	$\frac{1}{2}(5+\sqrt{29})$	70	13	-1
30	$11+2\sqrt{30}$			1
31	$1520+273\sqrt{31}$			1
32		17	3	1
33	$23+4\sqrt{33}$	23	4	1
34	$35+6\sqrt{34}$			1
35	$6+\sqrt{35}$			1
37	$6+\sqrt{37}$	6	1	-1
38	$37+6\sqrt{38}$			1
39	$25+4\sqrt{39}$			1
40		19	3	1
41	$32+5\sqrt{41}$	32	5	-1
42	$13+2\sqrt{42}$			1
43	$3482+531\sqrt{43}$			1
44		199	30	1
45		161	24	1
46	$24335+3588\sqrt{46}$			1
47	$48+7\sqrt{47}$			1
48		7	1	1

m	μ_0	x	y	D
50		7	1	−1
51	$50 + 7\sqrt{51}$			1
52		649	90	1
53	$\frac{1}{2}(7 + \sqrt{53})$	182	25	−1
54		485	66	1
55	$89 + 12\sqrt{55}$			1
56		15	2	1
57	$151 + 20\sqrt{57}$	151	20	1
58	$99 + 13\sqrt{58}$			−1
59	$530 + 69\sqrt{59}$			1
60		31	4	1
61	$\frac{1}{2}(39 + 5\sqrt{61})$	29718	3805	−1
62	$63 + 8\sqrt{62}$			1
63		8	1	1
65	$8 + \sqrt{65}$	8	1	−1
66	$65 + 8\sqrt{66}$			1
67	$48842 + 5967\sqrt{67}$			1
68		33	4	1
69	$\frac{1}{2}(25 + 3\sqrt{69})$	7775	936	1
70	$251 + 30\sqrt{70}$			1
71	$3480 + 413\sqrt{71}$			1
72		17	2	1
73	$1068 + 125\sqrt{73}$	1068	125	−1
74	$43 + 5\sqrt{74}$			−1
75		26	3	1
76		57799	6630	1
77	$\frac{1}{2}(9 + \sqrt{77})$	351	40	1
78	$53 + 6\sqrt{78}$			1
79	$80 + 9\sqrt{79}$			1
80		9	1	1
82	$9 + \sqrt{82}$			−1
83	$82 + 9\sqrt{83}$			1
84		55	6	1
85	$\frac{1}{2}(9 + \sqrt{85})$	378	41	−1
86	$10405 + 1122\sqrt{86}$			1
87	$28 + 3\sqrt{87}$			1
88		197	21	1
89	$500 + 53\sqrt{89}$	500	53	−1
90		19	2	1

m	μ_0	x	y	D
91	$1574 + 165\sqrt{91}$			1
92		1151	120	1
93	$\frac{1}{2}(29 + 3\sqrt{93})$	12151	1260	1
94	$2143295 + 221064\sqrt{94}$			1
95	$39 + 4\sqrt{95}$			1
96		49	5	1
97	$5604 + 569\sqrt{97}$	5604	569	-1
98		99	10	1
99		10	1	1
101	$10 + \sqrt{101}$	10	1	-1

Chronological Table

Mathematicians Prominent in the History of Number Theory or mentioned in the Text

Pythagoras	ca. 540 B.C.
Euclid	ca. 300 B.C.
Diophantos	ca. 250 (?)
Bachet de Méziriac	1581–1638
Marin Mersenne	1588–1648
Pierre de Fermat	1601–1665
John Pell	1610–1685
Leonhard Euler	1707–1783
Joseph Louis Lagrange	1736–1813
John Wilson	1741–1793
Adrien Marie Legendre	1752–1833
Karl Friedrich Gauss	1777–1855
Augustin-Louis Cauchy	1789–1857
Augustus Ferdinand Möbius	1790–1868
Gabriel Lamé	1795–1870
G. J. Jacobi	1804–1851
Peter Gustav Lejeune-Dirichlet	1805–1859
Ernst Eduard Kummer	1810–1893
Ferdinand Gotthold Eisenstein	1823–1852
Leopold Kronecker	1823–1891
Georg Friedrich Bernhard Riemann	1826–1866
Richard Dedekind	1831–1916
Adolph Hurwitz	1859–1919
David Hilbert	1862–1943
Axel Thue	1863–1922
Hermann Minkowski	1864–1909
Emil Artin	1898–1962

Bibliography

Cohn, H., *A Second Course in Number Theory*, Wiley, New York, 1962.

Gauss, K. F., *Disquisitiones Arithmeticae* (translated by A. A. Clarke), Yale University Press, New Haven, 1966.

Grosswald, E., *Topics from the Theory of Numbers*, Macmillan, New York, 1966.

Hardy, G. H., and Wright, E. M., *An Introduction to the Theory of Numbers*, 4th Ed., Oxford University Press, New York, 1960.

Nagell, T., *Introduction to Number Theory*, Wiley, New York, 1951.

Niven, I., and Zuckerman, H., *An Introduction to the Theory of Numbers*, Wiley, New York, 1960.

Ore, Oystein, *Number Theory and Its History*, McGraw–Hill, New York, 1948.

Pollard, H., *The Theory of Algebraic Numbers*, Carus Mathematical Monograph #9, Math. Assoc. Amer., 1950.

Sommer, J., *Vorlesungen über Zahlentheorie*, B. G. Teubner, Leipzig, 1907.

Vinogradov, I., *Introduction to the Theory of Numbers*, Pergamon, New York, 1955.

Weiss, E., *Algebraic Number Theory*, McGraw–Hill, New York, 1963.

List of Symbols

l.c.m. (a, b) least common multiple of a and b, 4

μ fundamental unit in $A(m)$ when $m > 0$, 97

$N(x)$ norm of $x \in Q(\sqrt{m})$, 84

$\Phi(n)$ group of units of Z_n, 15

φ Euler φ-function, 15

Q the field of rational numbers, 82

$Q(m)$ the least field containing Q and $\sqrt{m}$, 82

R^* nonzero elements of the ring R, 15

$R(x)$ ring of polynomials with coefficients in R, 66

$\sigma(n)$ the sum of the divisors of n, 24

$\omega = -\dfrac{1}{2} + -\dfrac{3}{2}$ a complex cube root of 1, 24

$[x]$ the largest integer less than or equal to x, 64

Z the ring of integers, 3

Z_n the ring of integers modulo n, 10

Index